POPULAR LECTURES ON MATHEMATICAL LOGIC

POPULAR LECTURES ON MATHEMATICAL LOGIC

Wang Hao

 VAN NOSTRAND REINHOLD COMPANY
NEW YORK   CINCINNATI   ATLANTA   DALLAS   SAN FRANCISCO
LONDON   TORONTO   MELBOURNE

 SCIENCE PRESS
Beijing, People's Republic of China

Van Nostrand Reinhold Company Regional Offices:
New York   Cincinnati   Atlanta   Dallas   SanFrancisco

Van Nostrand Reinhold Company International Offices:
London   Toronto   Melbourne

Library of Congress Catalog Card Number: 80–12062
ISBN: 0–442–23109–1

Manufactured in the United States of America

Published by Van Nostrand Reinhold Company
135 West 50th Street, New York, N.Y. 10020

Published simultaneously in Canada by Van Nostrand Reinhold Ltd.

15 14 13 12 11 10 9 8 7 6 5 4 3 2 1

Library of Congress Cataloging in Publication Data

Wang, Hao, 1921–
    Popular lectures on mathematical logic.

    Includes index.
    1. Logic, Symbolic and mathematical   I. Title.
QA9.W36   511.3   80–12062
ISBN 0–442–23109–1

PREFACE

In October 1977, I gave six extensive popular lectures on mathematical logic at the Chinese Academy of Science. In working over these lectures for publication, I have taken the liberty to express many of my immature opinions and to discuss, for the purpose of a moderate degree of completion, technical aspects on which my knowledge is very limited. Moreover, unexpected interruptions during the preparation of these manuscripts have had negative effects. Hence, there are inevitably many mistakes of one kind or another. Whenever diversified developments can only be illustrated by examples, I have naturally chosen samples familiar to me.

The problem of organization is not easy. Generally the lectures are independent of one another so that it is not necessary to read the chapters in sequential order. Occasionally some concepts and theorems are repeated in different parts with cross references. The three appendices are probably the most elementary part of the book. Chapters 3 and 4 also do not presuppose familiarity with mathematical logic. The first and last chapters are general summaries. The remaining four chapters are rather uneven: parts of them presuppose very little knowledge; other parts touch on complex matters only in an apparently simple manner.

The reader is advised not to spend too much time on any particular point that seems to present serious difficulties. Quite possibly some error is made in the text or there is a defect of leaving out necessary preliminaries in the presentation. Since the book attempts to interest readers with widely different degrees of mathematical training and to discuss too many topics in brief space, it is to be expected that various readers will find certain parts too elementary or too advanced.

v

CONTENTS

Preface v

1. ONE HUNDRED YEARS OF MATHEMATICAL LOGIC 1
2. FORMALIZATION AND THE AXIOMATIC METHOD 11
   2.1 Formal systems as special cases of axiom systems.... 11
   2.2 The predicate calculus or the first order logic..... 13
   2.3 Formal systems and formal thinking.............. 15
   2.4 First order and second order theories.............. 17
   2.5 Outline of Gödel's incompleteness theorems 19
   2.6 Background and breaking up of the proofs........ 21
   2.7 Undecidable mathematical statements 23
3. COMPUTERS 28
   3.1 Broad relevance 28
   3.2 Developing computer science 29
   3.3 Progress of computers 31
   3.4 Computers and the Chinese language 34
   3.5 Some examples of computer applications 35
   3.6 Unified college admission 37
   3.7 Proof of the four-color theorem 42
   3.8 Computer proof of theorems 45
4. PROBLEMS AND SOLUTIONS 49
   4.1 Problems as a driving force 49
   4.2 Problems in mathematical logic 52
   4.3 Some relatively transparent problems 55
   4.4 The Diophantine problem 58
   4.5 Euler paths and Hamilton paths 61
5. FIRST ORDER LOGIC 63
   5.1 Satisfiability and validity 63
   5.2 Reduction classes and the decision problem of first
       order logic 65
   5.3 The propositional logic 69
   5.4 Model theory 73
   5.5 The Löwenheim-Skolem theorem 77

5.6	Ultraproducts	80
5.7	Ramsey's theorem and indiscernibles	83
5.8	Other logics	86
5.9	Formalization and completeness	88
6.	COMPUTATION: THEORETICAL AND PRACTICABLE	94
6.1	Computation in polynomial time	94
6.2	The tautology problem and NP completeness	96
6.3	Examples of NP problems	100
6.4	The tautology problem	101
6.5	Polynomial time and feasibility	107
6.6	Decidable theories and unsolvable problems	108
6.7	Tiling problems	110
6.8	Recursion theory: degrees and hierarchies	112
7.	HOW MANY POINTS ON THE LINE?	119
7.1	Cantor and set theory	119
7.2	Finite set theory and type theory	121
7.3	Axiomatization of set theory	123
7.4	Hilbert's intervention	127
7.5	Constructible sets	128
7.6	Consistency of GCH	131
7.7	Constructibility	134
7.8	The continuum problem	135
7.9	Set theory since 1960	136
7.10	GCH and the relativity of cardinality	139
7.11	The method of forcing	141
7.12	Forcing in a simple setting	146
7.13	Nonconstructible sets	150
7.14	Independence of the CH	153
8.	UNIFICATIONS AND DIVERSIFICATIONS	156
8.1	Proof theory and Hilbert's program	156
8.2	Constructivism	161
8.3	The axiom of determinacy	162
8.4	Comments on the literature in mathematical logic	165
8.5	Hierarchies and unifications	166
Appendix A.	DOMINOES AND THE INFINITY LEMMA	170
1.	Some games of skill	170
2.	Thue sequences	179

3. The infinity lemma 182
4. A solitaire with dominoes (tiling problems) 186
5. The infinity lemma applied to dominoes 193
Appendix B. ALGORITHMS AND MACHINES 198
1. Numerical and nonnumerical algorithms 198
2. A programming prelude to abstract machines 202
3. Human computation and practical computers 206
4. A conceptual analysis of computations 208
5. Five contrasts concerning machines 214
Appendix C. ABSTRACT MACHINES 222
1. Finite-state machines 222
2. Turing machines 227
3. The P-machine (a program formulation of Turing
machines) 234
4. Unsolvable tiling problems 248
5. Tag systems and lag systems 257
INDEX 271

POPULAR
LECTURES ON
MATHEMATICAL
LOGIC

1. ONE HUNDRED YEARS OF MATHEMATICAL LOGIC

We begin by stating some of the central concepts and theorems in a historical and developmental context. Relations to other disciplines will also be briefly outlined. A number of the issues sketched in this chapter will be considered more closely in the other chapters and the appendices.

It is customary to trace back to Leibniz (1646—1716) for conceptual and fragmentary anticipations of modern logic. His notion of a theory of arrangements includes a calculus of identity and inclusion that is in the direction of Boolean algebra. He strove for an exact universal language of science, and looked for a calculus of reasoning so that arguments and disagreements can be settled by calculation.

In 1847, George Boole (1815—1864) published his *Mathematical analysis of logic* in which an algebra of logic is developed, now commonly known as Boolean algebra in mathematics. As logic, it corresponds to the propositional calculus, and has been extended to forms corresponding to the predicate calculus by C. S. Peirce and O. H. Mitchell in 1883; notable contributions along this tradition have been made by E. Schröder, L. Löwenheim, and Skolem.

It is convenient to date the beginning of mathematical logic back to shortly before 1880 when the predicate calculus (first order logic) and set theory were introduced[1]. Two directions of mathematics

---

1) There are two fairly long books on the history of formal logic: I. M. Bochenski, *A history of formal logic* (trans. I. Thomas), 1961; W. and M. Kneale, *The development of logic*, 1962. An outline of a more ambitious project to deal with logic in a broader sense is: Heinrich Scholz, *Abriss*, 1931, translated into English as *Concise history of logic*, 1961. There is a collection of original papers in mathematical logic from 1879 to 1931, all in or translated into English: *From Frege to Gödel*, ed. J. van Heijenoort, 1967. Collected (or selected) papers of Cantor, Hilbert, Brouwer, Skolem, Herbrand, Gentzen, etc. all have appeared.

1

played an important role in the background: the interest in the axiomatic method; and the concern with the foundations of analysis (in particular, real numbers and arbitrary functions of them). One main trend since about 1880 has been the interplay of the predicate calculus with set theory.

Another important trend was the study of the axiomatic method, both in getting and studying axiom systems for number theory, geometry, analysis, and set theory; and in a general study of the concept of formal systems as a sharpening of the concept of axiom systems with strong results on the limitations of formalization[1].

One achievement is an exact concept of theoretical computability which yields the discipline of recursion theory and gives a framework for introducing a theory of computation. This last aspect has so far produced little that is immediately useful for real computation, as the major results deal only with theoretical decidability and undecidability. The development of a theory of feasible computation remains at a rather primitive stage. What is most relevant to the advances in computers is rather the concern common to both areas for explicit formalization which makes mathematical logic a helpful discipline toward training experts in programming and especially in the design of programming languages[2]. A more obvious application is to the design of logical circuits for computers since these circuits correspond naturally to Boolean expressions or expressions in the propositional calculus. By the way, Boolean algebra and a number of simple algorithms are rather simple and can be taught in middle schools.

In recent years mathematical logic has found applications in number theory (e.g., the problem of deciding whether a Diophantine equation is solvable), algebra (e.g., the word problem for groups, and a proof of Artin's conjecture on p-adic fields), topology (notably proofs of consistency and independence), and an exact theory of infinitesimals. This last application uses a technique first introduced in giving a nonstandard model for the positive integers and touches on a broad philosophical question of intuitive reasoning and its rigid

---

1)   See Chapter 2.
2)   See Chapter 3.

cannot be defined in number theory and his plan of relative consistency proof did not work. He went on to draw the conclusion that in suitably strong formal systems there are undecidable propositions. Hence, the influence of Hibert's program is clear.

It must have been in 1930 when Gödel began to think about the continuum hypothesis and first heard about Hilbert's proposed outline of a proof of it. He felt that one should not build up the hierarchy in a constructive way and it is not necessary to do so for a proof of (relative) consistency. The ramified hierarchy came to his mind. By 1935 Gödel had obtained the concept of constructible sets, proved that the axioms of set theory (including the axiom of choice) hold for it, and conjectured that the continuum hypothesis also would hold. In the summer of 1938, Gödel extended his results to the extent familiar today: that the generalized continuum hypothesis cannot be disproved by the axioms of set theory, the first major result on the continuum problem since Cantor proposed it in 1878.

It follows from Gödel's incompleteness result that there are nonstandard models for any given formal system of number theory. For an extended period, Skolem had attempted to find nonstandard models of number theory. In 1933 he produced such a model by a new method which to a great extent anticipated the use of ultra-products about twenty years later.

In the area of proof theory and intuitionism, beyond the central incompleteness theorems of Gödel, Gentzen produced a consistency proof for number theory in 1936 which has since undergone various refinements and extensions. In 1932, Gödel put forth an interpretation (or translation) of classical number theory in intuitionistic number theory. In 1942, he found also an interpretation of intuitionistic number theory by means of primitive recursive functionals, which was published only in 1958. A good deal of work has been done since to extend Gödel's interpretation to deal with classical analysis. There has also been extensive study of inductive definitions and predicative (or constructivistic) analysis. Troelstra and others have tidied up somewhat formal studies of intuitionism.

Model theory has gradually emerged since about 1950. At first, results appeared to be elegant reformulation of familiar facts. But then interesting theorems and applications began to appear. In particular, it was also used in the study of large cardinals so that around 1960 Hanf numbers appeared and Scott proved that measurable cardinals yield nonconstructible sets. Results often mentioned as being impressive are Morley's theorem on categoricity in power, and applications to algebraic problems by Ax and Kochen.

The strongest impact on the further development of set theory, including its interaction with model theory, came from Cohen's independence proof of the continuum hypothesis in 1963. Apart from the importance of the result, the method used turned out to have a wide range of applicability[1]. For example, Solovay soon proved the consistency with dependent choice of the proposition that every set of reals is Lebesgue measurable. (He has to assume that there are inaccessible cardinals; it remains an open problem whether this stronger assumption can be avoided[2].) Apart from using Cohen's method to get various independence results, there has been a general upsurge of interest in other aspects of set theory. For example, there has been much work on the axiom of determinancy including its relation to large cardinals and their relation to descriptive set theory, by D. A. Martin, Solovay, H. Friedman, and others[3]. Jensen and others have examined more closely the structure of constructible sets. Partition properties and indiscernibles (tracing back to Ramsey's theorem of 1929 first introduced to deal with a simple case of the Entscheidungsproblem) have been extensively studied by Rowbottom, Silver and others. At present, the continuum problem remains the most challenging special problem; toward general progress, efforts are made to clarify the nature of large cardinals (especially those which require nonconstructible sets).

The concept of general recursive functions and Turing's computable functions have led to recursion theory and a number of mechanically unsolvable problems which are of particular interest in one way or another. In terms of the general theory, studies have been made of degrees of unsolvability, the arithmetical hierarchy, the hyperarithmetical hierarchy, and generalizations to admissible ordinals

1)   See Chapter 7.
2)   Robert M. Solovay, *Annals of math.*, vol. 92 (1970), pp. 1—56.
3)   A brief summary of some of the results is given in Chapter 8.

and admissible sets, which interact with set theory and model theory. Of the results on particular problems, the most famous are probably the unsolvability of the problem on the existence of solution to Diophantine equation (Hilbert's tenth problem) established in 1970 and that of the word problem for groups established in 1955.

Turing applied his machines to show that the decision problem of the predicate calculus is unsolvable through representing each machine by a sentence of the predicate calculus. The representation has the property that the machine will halt eventually when beginning with a blank tape if and only if the sentence has no model. Since no machine can decide for all machines whether it will halt on a blank tape, there is no algorithm to decide generally whether a sentence of the predicate calculus has a model. In 1962, this method was greatly refined so that it is sufficient to use only sentences of the surprisingly simple form $\forall x \exists y \forall z M x y z$ (the $\forall \exists \forall$ sentences). Hence, even the decision problem for the $\forall \exists \forall$ case is unsolvable[1].

In 1971 Cook was able to represent each Turing machine with a given input by a sentence of the propositional calculus in such a way that the machine can "guess at" the answer to the question on the input in "polynomial time" if and only if the sentence is satisfiable. Stated in technical terms, this says that the problem of deciding whether a Boolean expression is satisfiable is NP-complete, where NP stands for "nondeterministic polynomial." This result has generated wide interest because it leads quickly to the conclusion that many different classes of computational problems are equivalent in terms of decidability in polynomial time.

An assumption or thesis is that a class of problems is solvable by real or feasible computation if and only if we have an algorithm by which a question of length n in this class can be decided in time $P(n)$, where $P(n)$ is a polynomial with n as the only variable. Much effort has been devoted to finding out whether any of the many equivalent classes can be decided in polynomial time. The central question is commonly expressed as asking whether $P = NP$. But there are some disturbing features in this enterprise. First, polynomial time can be very big so that, for example, if $n^{10^{2^0}}$ steps is necessary

---

1)   See Chapter 5 and Appendix C.

to settle a problem of length *n*, the algorithm can hardly be called feasible. Second, the problem is wide open insofar as we do not even have much weaker negative results: e.g., we do not even know that the satisfiability of Boolean expressions is not decidable in quadratic time[1].

A natural idea of taking advantage of the emphasis on formalization in mathematical logic is to attempt a systematic approach to proving theorems by computers. Toward 1960, there was surprising initial success in this direction. But the project has so far not attracted enough people with a proper approach and further progress to date has not been significant, except perhaps for applications to the verification of computer programs which depend on relatively simple logical deductions. On the other hand, more *ad hoc* applications of computers to assist the proof of theorems have some impressive results, notably in the recent proof of the four color conjecture with essential use of computers[2].

---

1)   See Chapter 6.
2)   See Chapter 3.

2. FORMALIZATION AND THE AXIOMATIC METHOD

2.1 Formal systems as special cases of axiom systems

There is a natural and strict concept of formalization which has been made precise in mathematical logic in terms of a precise concept of mechanical procedures[1]. According to this concept, a formalized rule is an algorithm, i. e., a procedure that can be carried out mechanically. In particular, a formal system is an axiom system in which any proposed proof can be checked mechanically to determine whether it is indeed a proof. It is a nondeterministic or many-valued mechanical procedure for generating all its proofs. Hence, formal systems are the meeting ground of formalization and axiom systems. Not all axiom systems are formal systems, and formalization need not lead to axiomatization.

The axiomatic method is an orderly way of summarizing experience. For example, the sixth in Hilbert's list of mathematical problems proposed in 1900 speaks of treating in the same manner as with the foundations of geometry, by means of axioms, those physical sciences in which mathematics plays an important part. It is clear from a recent review[2] of the progress on this problem up to 1976 that the purpose is not to arrive at formal systems but rather to single out and examine certain basic assumptions.

The most widely known axiom system is that of Euclid for geometry. It has had great influence in Western science and educa-

---

1) This concept is discussed in Appendix B; compare *From math. to philosophy*, 1974, pp. 81—99, which includes also some contributions by Gödel.

2) See the recent volume *Mathematical developments arising from Hilbert problems,* ed. Felix E. Browder, Am. Math. Soc., 1976. The essay by A. S. Wightman in this volume deals with the sixth problem; formal systems are obviously far from the central concern of the works reported there on the study of physics by means of axioms. Compare also Hilbert's paper on axiomatic thinking, reprinted in vol. 3 of his *Abhandlungen* (particularly, the remarks on p. 379).

tion for about two thousand years. For hundreds of years, it has been taught in schools. It would be of interest to reflect on the familiar experience of learning it in school. It has not been taught as a formal system and, in fact, much is implicitly assumed and most deductions are made after a number of familiar propositions have already been derived. Pedagogically, there is a controversial question over how formal an axiom system for geometry should be used in teaching geometry.

During the 19th century, some of the implicit assumptions in geometry were explicitly stated for the first time, and axiom systems for other disciplines emerged: alternative geometries, the predicate calculus, the theory of natural numbers ("Peano arithmetic"), the theory of real numbers (treated by Dedekind as a special kind of ordered field); on the other hand, axioms for abstract structures such as groups, rings, fields also appeared and were studied in algebra. We find here a contrast of two different motivations: gener-·ality *versus* unique characterization.

For example, when groups (or fields) are studied, the important point is that there are many different kinds of group, so that theorems about them have wide general applications. When axioms for natural numbers or real numbers are formulated, the intention is to find enough axioms to capture exactly just these numbers. We speak of a group, a field, a Lie algebra, etc. but not the (set of) natural numbers, the field of real numbers, the universe of sets, etc. When we study groups, we are studying a class of mathematical structures which share the one common property of satisfying the axioms for groups, while in other cases we have in mind a particular structure. It is natural to classify groups and fields by adding additional properties, to move, for example, from fields to ordered fields or even to algebraically closed ordered fields. One would like to capture all particular structures in this way. But the experience so far is that we do not capture (at least not fruitfully, and that is the main issue) natural numbers or real numbers in any homogeneous way so that algebra does not include all of mathematics, even though it has applications in most branches of mathematics. For brevity we shall speak of the distinction as between abstract axiom systems and concrete axiom systems.

2.2 The predicate calculus or the first order logic

Traditionally mathematics does not make its method of reasoning or its language an object of study. Mathematical logic attempts to study these aspects mathematically by first rendering explicit the inferences and the languages used. At the center of the explicit formulations is the predicate calculus or the first order logic[1] which codifies explicitly obvious inferences such as substituting a for b when $a = b$, inferring q when p is true and p implies q, inferring $\exists xFx$ from Fa, etc. The concepts dealt with are sometimes called logical constants which include identity $(=)$; the propositional or Boolean connectives and $(\wedge)$, or $(\vee)$, not $(\neg)$, only if $(\supset)$; the quantifiers $\forall$ (for all) and $\exists$ (for some). Axioms and rules of inference are given, determining a formal system P which is complete with respect to the intended interpretation. Many alternative formulations[2] can be taken as the system P.

The formal system P is a concrete axiom system as far as it attempts to characterize the logical constants in an unequivocal fashion. It is yet different from systems for set theory and number theory because it is meant to deal with any relations and functions so that it has an abstract feature in the sense that a theorem of P is supposed to be true no matter what particular relations (e.g., the membership relation) and functions (e.g., addition and multiplication) are taken as the realization of the symbols in P standing for relations and functions. This notion of universally true is called valid and P is complete in the sense that all valid sentences of P are theorems of P. All ordinary axiom systems whether concrete or abstract contain implicitly P (or something like it) as a basic part. For example, even when the axioms of geometry or number theory are stated explicitly, we have to adjoin P to it to obtain a formal system because otherwise logical inferences have to remain informal.

---

1) Compare Chapter 5 for more extended considerations of first order logic and its descendent, model theory.

2) See section 5.9.

2.2.1 *The completeness of P.*

The formal system P is complete; i.e., all valid sentences are provable in $P^{1)}$.

We may further elaborate the connection between mechanical procedures and formal systems as follows: If more and more has been made explicit in the axioms of geometry by adding the axioms for order and the predicate calculus, how can we be sure that we ever reach a completely formal system? The answer is that we do have systems in which when proofs are written out in complete detail, every line is either an axiom as explicitly stated or a consequence of one or more previously proved lines according to some explicitly stated rule of inference. That a proposed formal proof is indeed a proof can be checked mechanically. We have here one reason for having a precise concept of mechanical procedures.

The notion of validity corresponds very well to the traditional idea that theorems of logic should be "true in all possible worlds." For example,

$$(B(x, y, z) \wedge B(x, w, y)) \supset B(x, w, z)$$

is an axiom of geometry since B is intended to stand for between, but it is not one of logic because if $B(x, y, z)$ is interpreted as y is half-way between x and z, it is no longer true. On the other hand,

$$\neg (B(x, y, z) \wedge \neg B(x, y, z))$$

is a logical truth because no matter what B, x, y, z are, the above is true.

An example of valid inference from ambiguous and dubious premises is:

All pleasures (M) are transitory (P)
Immortality (S) is a pleasure (M)
_____
$\therefore$ Immortality (S) is transitory (P)

The whole inference is logically valid because we can change M, P, S in any way we wish.

---

1)   This is considered in detail in Chapter 5.

Thus the predicate calculus is abstract in disregarding special properties of the relations and functions under study. This abstract feature is essential for the development of model theory in recent years. Although there are also results about concrete axiom systems, the chief application of model theory is to place and generalize basic concepts and facts of algebra within the framework of the predicate calculus (and its extensions). Sometimes model theory is said to be the union of logic with universal algebra. While ordinary branches of mathematics deal with numbers, functions and sets, theorems about first order logic deal not only with sets (as mathematical structures) but their relations to sentences.

For example, if a consequence T can be derived from a set A of sentences by P, then, since proofs using P are finite in length, T must be a consequence of a finite subset of A. Using 2.2.1 (the completeness of P), it is easy to deduce[1]:

2.2.2   *The compactness theorem for P.*

If every finite subset of A has a model, then A has a model.

From this, it is easy to deduce, for example: if every finite subgroup of a group G can be ordered, then G itself can be ordered. There are many other and less simple theorems in model theory[2]. An old example is the Löwenheim-Skolem theorem which says in the simplest case that if a sentence has a model, then it has a countable model. Here again the basic simplicity of a sentence yields a proof of the theorem quite directly. A more general form is[3]:

2.2.3   *The Löwenheim-Skolem theorem.*

Any formal system formulated in the framework of P has a countable model, if it has a model at all.

2.3   Formal systems and formal thinking

Even when we are dealing with a formal system, the process of selecting the right axioms to obtain the formal system certainly requires more than formal thinking; when we try to prove a theorem

---

1) Rather 2.2.2 is a generalization of 2.2.1; see discussions in Chapter 5.
2) Some of these results are listed or discussed in Chapter 5.
3) See section 5.5.

in a formal system, we do not try all possible proofs in a mechanical way and even after we find a proof by using our intuition, we do not generally take the trouble of writing down a formal proof. Rather the interesting uses of formal systems are considerations about formal systems which draw general conclusions about formal systems on account of their being formal. For example, the Löwenheim-Skolem theorem just stated, the Gödel incompleteness theorems to be considered below, etc. It is a familiar misconception to believe that to do mathematical logic is to be engaged primarily in formal thinking. The important point is rather to make precise the concept of formal and thereby be able to reason mathematically about formal systems. And this adds a new dimension to mathematics.

Admittedly there used to be people who were drawn to mathematical logic partly because they had an obsession with making everything entirely explicit and absolutely secure even for a mechanical intelligence. Formal deductions would seem to offer the solution. The ability to be so formal is useful in writing computer programs (including ones for installing programming languages) but in doing mathematical logic what is generally useful is seeing that formalization can be done rather than actually doing it.

We have mentioned a precise concept of formalization and formal system, and emphasized that axiom systems need not be formal systems. However, within mathematical logic, since one is much interested in formal systems, axiomatizable theories are commonly identified with formalizable (more explicitly, with recursively axiomatizable) theories. In fact, the concept of theory is given a syntactical meaning rather different from common usage.

This illustrates a general problem about specialized uses of common words. Insofar as the specialized uses are more precise and enable us to say more definite things, they are valuable for the advance of knowledge. But if we are thinking in a more general setting, mistaking the special uses for the broader and less definite ones can cause a good deal of confusion. For instance, when we contrast formal thinking with dialectical thinking, we have in mind something much broader than obtaining results which can be written out in given formal systems in the sense we have explained. Rather it corresponds more or less to abstract thinking or, more broadly, it is taken to be any thinking that does not capture the real situation in its full concreteness. If we take formal thinking in such a broad sense, as one does sometimes, we might even say paradoxically that at each moment we are only capable of formal thinking and the essence of dialectics is to realize this and strive for better approximations to the whole real situation all the time. In fact, this would appear to be one possible interpretation of the following famous passage[1]:

"We cannot imagine, express, measure, depict movement without interrupting continuity, without simplifying, coarsening, dismembering, strangling that which is living. The representation of movement by means of thought always makes coarse, kills, ... and not only by means of thought, but also by sense-perception, and not only of movement, but of every concept.

And in that lies the *essence* of dialectics.

And precisely this *essence* is expressed by the formula: the unity, identity of opposites."

2.4   First order and second order theories

Formal systems of number theory and set theory have been much studied. But they are not the natural systems which people first came up with. The original systems included arbitrary sets of integers and arbitrary "definite" properties without corresponding axioms for them; similarly arbitrary (Dedekind) cuts in the axioms for real numbers and arbitrary point sets in the continuity axiom for geometry were included at first. These are now called second order theories because they bring in informally sets of the first order objects directly under attention. Harking back to the tradition of type theory, these sets are thought of as second order objects.

To illustrate this distinction, let us consider the Peano arithmetic PAII as commonly stated in mathematical textbooks. The primitive concepts are zero and successor. The axioms are:

---

1)   V. I. Lenin, *Philosophical notebooks* (Collected works, vol. 38), pp. 259—260.

PA1. $0 \neq x + 1$.

PA2. $x \neq y \supset x + 1 \neq y + 1$.

PA3II. For every set C, if $0 \in C$ and $\forall x \ (x \in C \supset x + 1 \in C)$, then $\forall x \ (x \in C)$.

This is known to be "categorical": i.e., any two models are isomorphic. By PA1, each model has a beginning; correlate them with each other. And then proceed to make correlations by adding 1. By PA2, we do not return to an earlier one. By PA3II, we reach every object in each model; otherwise we would have a set C such that $0 \in C$, $\forall x (x \in C \supset x + 1 \in C)$ but $\exists x \neg (x \in C)$.

This familiar argument works only because we do not really have a formal system. PA3II is out of place because we have no axioms governing the membership relation ($\in$) and the set variable C. To remedy this, we can either adjoin axioms for sets or alternatively use properties of natural numbers expressible in the given system and confine ourselves to sets thus expressible. In either case, we have Gödel's theorem, according to which formal systems of this kind are not and cannot be complete or categorical.

To obtain the formal system PA for the first order Peano arithmetic, we include the predicate calculus P and replace PA3II by recursive definitions for addition and multiplication and an infinite set of induction axioms one for each formula $F(x)$ in the language of PA[1]:

PA3.   For every formula $F(x)$ of PA, if $F(0)$ and

$$\forall x(F(x) \supset F(x + 1)), \text{ then } \forall x F(x).$$

PA4. $x + 0 = x, x + (y + 1) = (x + y) + 1$.

PA5. $x \cdot 0 = 0, x \cdot (y + 1) = (x \cdot y) + x$.

It should be noted that with the formulation of the predicate calculus, we automatically get a natural collection of properties (of natural numbers in the particular case of PA) which are expressible in the language of a first order theory (in this case PA). This is one of the most useful concepts in mathematical logic (we shall

---

1) For a more extended discussion of the Peano axioms, see *J. symbolic logic*, vol. 22 (1957), pp. 145—157; Dedekind's letter of 1890 (mentioned in Chapter 1) was published in this paper for the first time.

later on discuss its use in set theory). There is a derived notion of first order definable sets (also perhaps from parameters which for example may range over a collection of sets). In the case of PA, the first order definable sets (i.e., the sets determined by the properties expressible in PA) are commonly called the arithmetical sets.

2.5 Outline of Gödel's incompleteness theorems

2.5.1 *Gödel's first incompleteness theorem.*

A method is given of constructing, for any formal (formalized axiom) system S of mathematics, a question of number theory undecidable in the system.

This is a general result which depends essentially only on three conditions: (a) The axiom system S is truly formal; (b) The system S is rich enough for developing a moderate amount of number theory; (c) S is consistent[1]. Under these conditions, Gödel shows that the system S contains a statement F, in fact, of the simple form $\forall x G(x)$, where G is a decidable or recursive predicate, such that neither F nor $\neg F$ is a theorem of S. More specifically, $G(0)$, $G(1)$,... all are provable in S, but neither $\forall x G(x)$ nor $\neg \forall x G(x)$ is. Hence, in particular, there is some (nonstandard) model of S in which $\neg \forall x G(x)$, as well as $G(0)$, $G(1)$,..., all are true. In such a model there must be some "unnatural numbers" k, different from 0, 1, ... such that $\neg G(k)$ is true. Hence, S is not categorical.

---

1) Strictly speaking, a property stronger than consistency is needed: ω-consistency which requires that there is no formula $G(x)$ such that $G(0)$, $G(1)$, etc. and also $\neg \forall x G(x)$ are all theorems of S. In *J. symbolic logic*, vol. 1 (1936), pp. 87—91, J. B. Rosser modifies Gödel's original proof to make condition (c) sufficient. In place of the sentence (3) given below in the text, Rosser uses essentially a sentence:

(R) (R) is not provable in S by a proof such that there is no shorter proof of $\neg$ (R) in S; or, I am not provable in S by a proof shorter than a refutation of me.

This has the intuitive effect that if (R) is provable in S, it is also refutable in S and, therefore, S is inconsistent. In terms of the notation introduced later in section 2.6 of the text and using $d(x)$ for the sequence representing the negation of the formula represented by x, we can express Rosser's sentence as:

$$\forall x \neg [Prov(x, s(p, p)) \land \forall y (y \leqslant x \supset \neg Prov[y, d(s(p, p))])].$$

First we give an outline which is perhaps helpful only to those who are already familiar with the more formal proofs. We shall then return to a somewhat more detailed discussion of the crucial points. Here is the outline:

(1)   (1) is false.   (I am false.)
(2)   (2) is not provable.   (I am not provable.)
(3)   I am not provable in S.

It is easy to see that (1) and (2) lead to contradictions.

If (1) is true, i.e., ''(1) is false'' is true, then (1) is false.

If (1) is false, then ''(1) is false'' is true, i.e., (1) is true.

If (2) is provable, then (2) is true, i.e., ''(2) is not provable'' is true. Hence, (2) is not provable. If (2) is not provable, then ''(2) is not provable'' is true, and we have thus proved (2)[1].

But (3) is different. The second half of the argument for (2) breaks down, because it is possible that ''(3) is not provable in S'' is true but not provable in S. This is indeed what happens with S and a representation of (3) in S. That (3) can be expressed in S depends on a representation of formulas by numbers and properties of formulas by properties of numbers.

In particular, the statement that ''S is consistent'', i.e., for no F, are both F and $\neg F$ theorems of S, can also be expressed in S and is in fact not provable in S. In fact, this is a consequence of the proof of 2.5.1. The first half of 2.5.1 shows that if S is consistent then the sentence (call it p) expressing (3) in S is not provable in S. This argument can be formalized in S to give a conditional theorem of S: if $Con(S)$ (the sentence of S expressing that S is consistent), then p. Hence, if $Con(S)$ were a theorem of S, p would be one also, contradicting the first half of 2.5.1. Hence, we arrive at:

2.5.2 *Gödel's second incompleteness theorem.*

No classical formal system of mathematics can prove its own consistency. Or, the consistency of a system S satisfying (a), (b), (c) mentioned above has a natural representation by a sentence $Con(S)$ in S but $Con(S)$ cannot be a theorem of S.

---

1)   Compare Shen Yu-ting, *J. symbolic logic*, vol. 20 (1955), pp. 119—120.

2.6   Background and breaking up of the proofs[1]

As Gödel began his research shortly before 1930, a most central project is to prove the consistency of classical analysis which may be viewed as an extension of PA by adding sets of numbers and axioms governing them. We simply take PAII as described above and add axioms for the classes:

CA1 to CA3. Same as PA1, PA2, PA3II.

CA4. Comprehension. For any formula $\phi(x)$ of this new language, $\exists C \forall x (x \in C \equiv \phi(x))$.

CA5. Extensionality. If $\forall x (x \in C \equiv x \in X)$ then $C = X$.

We shall disregard the more or less routine reformulation of the predicate calculus needed for the class variables.

The crucial new axiom (schema) is CA4. The task Gödel initially set himself in the summer of 1930 was to derive the consistency of CA from the truth of PA or rather some variant PA^+ of it including other equally elementary concepts which we now know can be defined in PA. In other words, he was quite willing to assume the consistency of extensions of PA which add further true statements just about natural numbers and similarly elementary objects. It should be observed that at that time it was not known whether PA or some other extension of it might not be a complete formal system. In any case, Gödel permitted himself a flexible framework to begin with. His idea is to represent each set in CA (or, roughly speaking, each real number) by its definition given in CA4 so that if C corresponds to $\phi(x)$, then it can be viewed as determined by the subset of $\{\phi(0), \phi(1), \dots\}$ which consists of all and only the true ones for the particular $\phi(x)$ in question. But ϕ can be very complex especially when it involves quantifiers over sets. Even if we disregard for the moment the more complex ϕ's, we must be able to deal with at least those ϕ's which are arithmetical predicates, i.e., involve only the numbers. To do this, we need the notion of an arithmetical or number-theoretic truth.

---

1)   A careful proof of 2.5.2 was worked out for the first time in D. Hilbert and P. Bernays, *Grundlagen der Mathematik*, vol. II, 1939.

Let us now look at the language of PA$^+$. Gödel's earlier idea was to represent the symbols by numbers. A formula is then a sequence of numbers, and a proof is a sequence of sequences of numbers, which are again objects of PA$^+$. Suppose now we can find a formula $T(a)$ *in* PA$^+$ so that $T(q)$ is true if and only if q is a sequence of numbers representing a sentence of PA$^+$ which is true. Gödel was able to see that one would get a contradiction[1].

For this purpose, he introduced a substitution function $s(x,y)$ such that if x is a sequence representing a formula $F(a)$ with the free variable a, then $s(x,y)$ is the sequence representing the $F(y)$, i.e., the result of substituting y for a. This is a sophisticated construction but the function s is quite elementary and it is combinatorial in nature. After all it is easy to enumerate the sequences of numbers in some simple way and it is easy to single out those which represent formulas with a single free variable a. It is also a simple matter to locate those sentences obtained from such formulas $F(a)$ by substituting specific sequences of numbers for a.

Once such a function $s(x,y)$ is available, we are able to make the kind of self reference as in the statements (1), (2), (3) above, with, for example, (1) saying that (1) is not true. We have supposed that we have in PA$^+$ a truth predicate $T(a)$ for PA$^+$. Let us now consider the formula $\neg T\ (s(a,a))$ of PA$^+$. Since it is a formula of PA$^+$, it is represented by a sequence of numbers, say p. Consider now the sentence $\neg T(s(p,p))$ which is, say, represented by the sequence of numbers q. But according to the meaning of $s(x,y)$, we have: $q = s(p,p)$. Hence, $\neg T(s(p,p)) \equiv \neg T(q)$. But $\neg T(q)$ says that the sentence $\neg T(s(p,p))$ is not true. Hence, $\neg T(q)$ says of itself that it is not true. The conclusion is, therefore, that no such predicate T can be found in PA$^+$.

From this, Gödel saw that his approach to proving the relative consistency of CA did not work. He went on to consider what would happen if we move from truth to provability in PA$^+$. After all PA$^+$ is supposed to be a formal system. Therefore, the relation *Prov* (a, b) between a sequence of sequences of numbers a and a

1) For a discussion by Gödel of the relation of his theorems to the paradoxes, see §7 of his 1934 notes, Martin Davis, *The undecidable*, pp. 63—65.

sequence of numbers b such that a is a proof of b in PA^+, must be a quite elementary relation and therefore available in PA^+. If we repeat the argument of the preceding paragraph, we get into a different situation.

Let p be the sequence of numbers representing the formula $\forall x \neg Prov(x, s(a, a))$ of PA^+, and q be the one representing the sentence $\forall x \neg Prov(x, s(p, p))$ of PA^+. We have, as before, the situation that $\forall x \neg Prov(x, q)$ says of itself that it is not provable in PA^+. But this time we do not get a contradiction because it was an open question whether PA^+ was complete. In fact, the conclusion is rather that PA^+ is not complete because $\forall x \neg Prov(x, q)$ is not provable in PA^+ and similarly its negation is also not provable in PA^+ if PA^+ is at all a reasonable system.

This result was presented by Gödel in September, 1930 at a meeting[1]. Shortly afterwards, he sharpened the result to the extent of using essentially PA rather than PA^+ and deduced his second theorem also[2].

We shall not discuss here the various effects these two theorems have had over later developments but only consider the question of finding undecidable sentences of PA which are more like ordinary mathematical statements (e.g., not mentioning proofs and provability).

2.7 Undecidable mathematical statements

For many years attempts have been made to obtain results showing ordinary number-theoretic problems to be undecidable in systems like PA. For example, if Fermat's conjecture is undecidable in PA, then it is true because if it were false, there would be a numerical counterexample which could certainly be verified in PA. In other words, if it is false, it is provably false in PA.

---

1) This was a meeting at Königsberg where Gödel had discussions with J. von Neumann on Gödel's new discovery.

2) The famous 1931 paper is reprinted in *From Frege to Gödel* in English translation with notes added by Gödel. The 1934 lectures are reprinted in *The undecidable* with a postscript by Gödel.

There has been no success so far in finding any of the familiar open problems of number theory to be undecidable in PA. But last year some true statements on partition properties have been shown to be unprovable in PA. After these results came out in 1977 it became clear that, by some results already available in 1972, a more familiar statement (Ramsey's theorem) is not provable in a natural conservative extension of PA (i.e., an extension of PA in which there are no new theorems dealing only with objects of PA).

The extension PPA (predicative extension of PA) can be briefly described as the system obtained from CA by weakening the comprehension axiom CA4 to require that $\phi(x)$ is arithmetical (or that sets can be introduced only by formulas of PA). It is essentially the same system as PA except that we have in it a direct way of speaking about sets which are only the ones defined by arithmetical predicates.

This is a convenient place to introduce Ramsey's theorem and the arithmetical hierarchy. The arithmetical sets and predicates can be classified in a natural manner according to a measure of complexity. We can put all quantifiers at the beginning of a predicate (called the prenex normal form) and combine consecutive quantifiers of the same kind (namely, several $\forall$'s or several $\exists$'s) into one . Hence, each predicate is of the form $Q_1x_1Q_2x_2Q_1x_3Q_2x_4\ldots F(x_1, x_2,\ldots)$, where F contains no more quantifiers and Q_1 is either $\forall$ or $\exists$ (and then Q_2 is $\exists$ or $\forall$). A predicate is said to be $\Sigma_n(\Pi_n)$ if it can be expressed in the above form with n quantifiers beginning with $\exists$ ($\forall$). It is said to be Δ_n if it can be expressed both as Σ_n and as Π_n. This classification yields a hierarchy because if $m > n$, any Σ_n or Π_n predicate can be expressed as Σ_m and Π_m, but there are predicates which are Σ_m or Π_m but are not Σ_n or Π_n.

Let $[A]^n$ be the set of all n-element subsets of A. If A is a set of PPA, $[A]^n$ is also a set of PPA. A partition X of $[A]^n$ is generally understood as a separation of $[A]^n$ into a finite number of mutually exclusive subsets. If A is a set of PPA, every arithmetical partition X is also a set of PPA (or can be coded by a set of PPA). We can state Ramsey's theorem in several ways[1]:

---

1)   More is said about Ramsey's theorem in Chapter 5.

2.7.1   *Ramsey's theorem.*

Given any infinite set A and any n and a partition X of $[A]^n$, there exists an infinite subset B of A such that $[B]^n$ is contained in one part of the partition (i.e., if $[A]^n = Y_1 \cup \cdots \cup Y_k$, then $[B]^n \subseteq Y_i$, for some i, $1 \leqslant i \leqslant k$).

This is a statement of PPA essentially of the form $\forall X \exists Y$ $F(X, Y)$. Sets like B are often said to be homogeneous for the partition X. Hence, 2.7.1 can also be stated briefly as follows: For every partition of $[A]^n$, with A infinite, there is an infinite homogeneous subset of A. Another way of stating 2.7.1 is: if $X: [A]^n \to k$, where A is infinite and n, k are finite, then A has an infinite homogeneous subset for X. Various generalizations of 2.7.1 have been studied for the last fifteen years or so and a general notation has evolved. The partition relation $\beta \to (\alpha)_\gamma^m$, where α, β, γ are ordinals and m is finite, holds iff, whenever A is a set of order type β and $X: [A]^m \to \gamma$, then there is a homogeneous subset B of A for X which is of order type α. In this notation, 2.7.1 is simply: $\omega \to (\omega)_k^n$, where ω is the order type of the natural numbers.

Back in 1972, Jockusch published the result[1]:

2.7.2   Let N be the set of natural numbers. For every $n \geqslant 2$, there exists a (recursive) partition X of $[N]^n$ into two parts such that there is no $\sum_n$ homogeneous set for X.

Several people have noticed that a short argument would lead from 2.7.2 to the unprovability of 2.7.1 in PPA. One way to do this is to use the following lemma, where $\sum_n(X)$ means $\sum_n$ relative to X:

2.7.3   If PPA $\vdash \forall X \exists Y F(X, Y)$, then, for some k,

PPA $\vdash \forall X \exists Y \in \sum_k(X)(F(X, Y))$.

This almost certainly is a known theorem in proof theory. But it can also be proved quite directly using model theory[2]. Given

---

1)   Carl G. Jockusch, *J. symbolic logic*, vol. 37 (1972), pp. 268—280. See, in particular, Theorem 5.1 on p. 275.

2)   Jockusch has communicated to me the following proof of 2.7.3. Let $G_k(X)$ be the formula of PPA saying $\exists Y \in \sum_k(X) (F(X, Y))$. Let T be the theory whose axioms are those of PA with the induction axiom relativized to C together with all sentences $\neg G_k(C)$, $k = 0$, 1, 2, $\cdots$, where C is a new constant symbol for sets. Assume that for no k is $\forall X G_k(X)$ provable in PPA. Then since $F_k(X)$ implies $F_{k+1}(X)$ in PPA, T is consistent. Let M be a model

2.7.2 and 2.7.3 it follows directly that 2.7.1 is not provable in PPA. All we have to do is to take $F(X, Y)$ as saying that Y is homogeneous for the partition X of $[N]^n$, where X is taken as ranging over all (recursive) partitions of $[N]^2$, $[N]^3$, etc. If 2.7.1 were provable in PPA, there would be by 2.7.3 some fixed k such that for every n and every partition X of $[N]^n$, there is a $\sum_k(X)$ homogeneous set for X. But 2.7.2 says that there can be no such k. Hence 2.7.1 is not provable in PPA.

2. 7. 4   The Ramsey theorem 2.7.1 is not provable in PPA.

Reflecting on some joint work with Kilby, Paris first arrived at some unprovable true statements in PA which, unlike Gödel's original undecidable statements, do not require coding of logical concepts such as provability. With the help of Harrington, more attractive statements are found. We state only some of these results[1]. It seems natural to view these as a refinement of 2.7.4 in order to work with PA rather than PPA while paying the price of using a less familiar statement. Let a finite set of natural numbers be said to be relatively large if it has at least as many members as its smallest element. For example, if a finite set X contains 999 as its smallest member, then it must have at least 999 members to qualify as a relatively large set. To accommodate the use of this

---

of T. Expand M to a model M' of PPA by letting the sets of M' be all subsets of M first order definable in M relative to the constant C. Then M' does not satisfy $\exists Y F(C, Y)$ and therefore, $\forall X \exists Y F(X, Y)$ is not a theorem of PPA.

Jockusch also mentioned a previous proof of Solovay using a different but related lemma:

2.7.3′   If $\forall n \exists X F(n, X)$ is a theorem of PPA and contains no other set variables, then, for some k, PA $\vdash \forall n \exists X \in \sum_k (F(n, X))$. Here it is necessary (and possible) to turn the last formula into one of PA, eliminating the set variable X. The proof of 2.7.3′ is similar to that of 2.7.3. Given 2.7.3′, $F(n, X)$ can be taken as saying that X is infinite and homogeneous for the n-th recursive partition of $[m]^k$ into two sets for some k. Since $\forall n \exists X F(n, X)$ is then a consequence of 2.7.1, if 2.7.1 is provable in PPA, then 2.7.3′ yields some k such that $\forall n \exists X \in \sum_k (F(n, X))$, contradicting 2.7.2. Hence, 2.7.1 is not provable in PPA.

1)   See Chapter D8 of *Handbook of math. logic*. References to previous papers of J. Paris and L. Kirby are given there.

concept the notation $j \rightarrow (k)_n^m$ is modified to $j \xrightarrow{*} (k)_n^m$ to require that the homogeneous set with k members be relatively large.

2.7.5   The following statement of PA is not a theorem of PA:

$$(4) \qquad \forall k \, \forall m \, \forall n \, \exists j (j \xrightarrow{*} (k)_n^m).$$

More explicitly, (4) says that for any k, m, n, we can find some large enough j such that there is a set A of natural numbers with j members so that for every partition X of $[j]^m$ into n pieces, there is a relatively large subset of A with k members which is homogeneous for X. If we delete * from (4) or equivalently delete "relatively large" from the longer statement, then the result (4′) is just the finite Ramsey theorem. In fact, (4′) is a theorem of PA. But the natural proof of (4) uses the (infinite) Ramsey theorem 2.7.1 which cannot be stated in PA.

An interesting aspect of these results is their relation to the speed in which the value of a recursive function increases with the argument value. It is well known that while all partial recursive functions can be described simply, for each formal system S we can find a description which determines a total function but cannot be proved to do so in S. The statement that a particular description defines a total function is of the form $\forall m \exists n F(m, n)$ where $F(m, n)$ is a simple (e. g., primitive recursive) predicate. Let $f(n)$ be the least j such that $j \xrightarrow{*} (n + 1)_n^n$.

2.7.6   If g is a description of a recursive function and PA ⊢ "g is total", then, for all sufficiently large n, $g(n) < f(n)$.

Hence, $f(n)$ dominates all provably recursive functions $q(n)$ in PA, and $f(n)$ increases faster than every such $g(n)$. Clearly (4) determines a recursive function $j = f(k, m, n)$, but it increases so fast that no function g provably recursive in PA can have the property that $g(m, n, k) \geqslant f(m, n, k)$ for all m, n, and k.

3. COMPUTERS

3.1 Broad relevance

Unrealistically speaking, everything a computer can do could be done by human beings, since, given enough time, perseverance, and good health, a moderately intelligent person can certainly carry out the routine, even if sometimes complex, operations delegated to a computer. This realization is of interest for certain special theoretical considerations of idealized situations, but is quite misleading otherwise. To begin with, computers certainly help to save mental labor of a tiresome sort. Moreover, in practice they enable us to perform elaborate computations which would not be doable without them. In addition, their speed makes an essential difference when it is necessary to get something done in a limited amount of time (in particular, in "real time") such as in weather forecasting and economic decisions.

We have had occasion to discuss the special sense of formalization as related to the concept of mechanical procedure. It is easy to see that computers need not work exclusively with numerical calculations but can deal with all sorts of formalized procedures in and outside of mathematics. Numerical calculations are only a peculiarly obvious and traditional form of explicit procedures which can be handled by machines dealing primarily with symbols. Hence, it is customary to speak of information or data processing rather than computation. For example, computers are widely used in the operation of telephone exchanges, airplane reservations, printing, automation, the study of organic compounds, etc.

The broad range of computer applications introduces a new dimension of thought into many human activities. Apart from situations where computers are obviously helpful, one is faced with the difficult problem of finding new applications of computers. The essential task is to make explicit what one knows only implicitly. This can be very difficult indeed: for example, to recognize handwriting or to transcribe speech into written language has had little success so far.

The social consequences of computers, if not carefully attended to, can also include many unnecessary ones which are drastically bad. For example, key-punching as a specialized job is one of the most meaningless types of work there is[1]. Professional computer programmers often wish to leave the field after a few years. As with earlier machines, there is the phenomenon of people becoming the slaves of machines. Computers have also been used to confuse issues and divert attention from real social and scientific problems[2]. Of course, the abuse of science and technology is nothing new and the important thing is exactly to develop the positive force of science and technology while avoiding the unnecessary abuses as more and more progress is made.

3.2 Developing computer science

Computers as are familiar today have undergone a rapid development over only the short period of less than forty years even though the basic conceptions were introduced by Charles Babbage in 1833 in a practical form and by A. M. Turing in 1936 in an idealized form. The pace of progress has been truly amazing. Roughly speaking, the failure rate and the size of a basic functioning unit (a flip-flop) has been reduced and the speed has been increased by ten thousand times, the power consumption per unit has been reduced by a thousand times, and the cost per unit by one hundred times. Many new applications have been introduced.

There are intricate interactions between computers and other branches of science and technology. To make computers, the basic components not only depend centrally on electronics and in particular semiconductor physics, but also need a high level maturity of optics, materials science, etc. The varied applications of computers to different branches of science usually require familiarity with the particular science. Hence, one easily gets the impression that computer science is a motley of diverse things, requiring knowledge of many areas which is too much to ask from one person. The actual situation is rather different. Like other sciences, there is specializa-

---

1) See Harry Braverman, *Labor and monopoly capital,* 1974.
2) See Joseph Weizenbaum, *Computer power and human reason,* 1976.

tion within computer science. Moreover, it is possible to locate a hard core of computer science which interacts with the other disciplines only on the fringe. Both for the construction and for the application of computers, the knowledge needed of the other disciplines is generally limited to what is relevant: what existing principles and products are useful for the improvement of components or what problems are amenable to using the computers. It is usually desirable to have specialists in these other disciplines to keep computers in mind with an eye to the possibility of applying their results to improve computers and applying computers to help their own work.

A designer (probably a structured group of people) of the hardware of a computer needs knowledge of what components are available and his central problem is to find a good way of putting together a good selection of available components. It is of course desirable to plan both the hardware and the software together, insofar as that is feasible. In any case, there is an important part of computer science that is primarily concerned with thinking about algorithms and procedures especially when one comes to think about designing programming systems and finding new applications.

The point is that partly because computer science is a young subject and partly because it has a wide range, there is a good deal of room for innovations by people who do not have long years of preparation, especially if they tend to be comfortable with combinatorial or algorithmic problems in the broad sense of these terms. This is quite different from the situation with older, more concentrated branches of science and technology. In a sense, computer science is more elementary because much of its content is more explicit so that the complexities can be completely unravelled with enough effort to attain ultimate local precision.

For the same reason much of the central parts of computer science can be taught widely at an early age. For example, programming appears mysterious and formidable at first sight but the rudiments of it can be learned quite quickly by most students in secondary schools. Similarly, Boolean algebra, Turing machines and finite automata all are elementary and simple in the same way.

We have mentioned the surprisingly rapid development of computers over a short period. There is another aspect where the development has been disappointing, namely in the area known as "artificial intelligence". This is an area[1] in which over the years there have been many honest wild dreams (such as Turing's imitation game[2]) and many dishonest wild claims. It is a rather delicate matter to locate the proper balance so that while serious bold approaches to difficult new tasks are being encouraged, deceptions are prevented and unnecessary waste is held at a low level. It is necessary to make careful preparations to break up the difficulties into manageable parts and yet to avoid the other extreme of not beginning experiments on computers until every step is thought out in advance.

For the purpose of catching up with advanced research in a given subject, there is the problem of making proper use of the tradition of the subject and the tradition of the society insofar as it bears on the pursuit in question. The tradition of a subject generally contains good elements and bad elements; it is sometimes more autonomous, sometimes less. For example, in many cases sincere close cooperation of different workers cn a problem can greatly increase the effect. Too much prejudice against applied research or against "pure" research can both be harmful for rapid genuine scientific progress.

It is often possible to make interesting new discoveries in borderline regions because being on a borderline implies that it is a relatively new area which has not yet been examined thoroughly. There is, however, a familiar danger here. For example, there has been a large amount of publications dealing with rather arbitrary variations of idealized models of computers. Most of the concepts and results are neither interesting mathematics nor bear significantly on the operations of actual computers. Yet the authors are able to tell the theoretical people that they are doing practical work and tell the practical people that they are doing theoretical work.

3.3 Progress of computers[3]

Consider first the simple problem of doing addition of two or

---

1)   For extended discussions see *From math. to philosophy*, 1974, pp. 280—328.

2)   A. M. Turing, Computing machinery and intelligence, *Mind*, vol. 59 (1950), pp. 433—460.

3)   For a general discussion of different aspects of computers, see the special issue of *Science*, March 18, 1977, vol. 195, no. 4283.

more numbers on an abacus. Numbers are represented by counters in the familiar and natural manner. It is easy to imagine adjoining an auxiliary device to it so that carrying is done automatically whenever the counters on a column add up to ten or more. The result would then be like an adding machine except for the fact that numbers are represented in different forms and intermediate results are displayed on the abacus. Conceptually it is not a big step if we include also multiplication and other operations and have a machine perform a succession of fixed operations. A new element is rather the inclusion of choices or decisions in the form of a branching (or jump or conditional transfer) operation. For example, continue to do one thing as long as the index is less than 50 and then do something else.

During the Second World War an "automatic sequenced controlled calculator" was constructed with relays which were familiar in telephone equipment. Instructions were "written" on a plugboard while the data were stored inside. Shortly afterwards the idea of stored programs appeared with a great gain in flexibility. We can, for example, introduce instructions to modify other instructions at suitable junctures. These "stored program computers" are regarded as the basic form of modern computers.

A computer consists of five basic units: the central control unit (CPU), the input (I), the output (O), the arithmetic-logical unit, the memory (or store). The input transmits data and instructions to the store. The control unit is to locate in the store the (next) instruction to be carried out, decode it, and, depending on the result of the decoding, either (a) carry out a "computation" in the arithmetic-logical unit, or (b) put the result of a "computation" at an appropriate location in the store, or (c) determine the next instruction to be carried out by examining the content of some register, or (d) input a word to the store, or (e) output a word from the store. Except in the case of (c), the control simply goes to the next instruction on the list of instructions in the store.

As computers continue to develop, there are many improvements. Apart from software advances, the components get faster and smaller. In the mode of operation, parallelism is introduced to speed up performance. While one operation is being performed, it is sometimes possible to work on the next instruction as well. By the use of buffering and channel independence, one can attain input-output concurrency so that computation can go on at the same time certain material goes through the input or the output. Moreover, since input and output are generally slower, the central computer can work on other programs while one program is waiting for more inputs or in the middle of sending out outputs. This is called multiprogramming. Similarly, one can also have several processing units linked together in order to carry out simultaneously different parts of a program. This is called multiprocessing.

Time-sharing is a more explicit form of multiprogramming under which each user has the impression of using the whole computer, communicating directly with it through a terminal. The central control undertakes the task of distributing and scheduling the available space and time in the computer for programs of different users. Computer network is a more explicit form of multiprocessing. Different computers sometimes located at far away places are connected together so that a user can utilize the combined power of these machines. Of course all these advantages are obtained at great expense in terms of the great effort needed to bring about the functioning systems and sometimes added complexities to the users.

On the side of the software, the main problem is to communicate between people and computers. What a computer deals with directly are sequences of binary digits (0 or 1, yes or no, on or off). The problem of designing programming languages is to combine exactness with flexibility, to get close to natural language while translatable into the rigid machine language. The primary goal is to introduce languages which are easy to use, under the limitation that they must be precise enough to be transformable into manipulations with strings of binary digits. The best known programming languages are probably the different editions of ALGOL and FORTRAN. Other examples are COBOL, BASIC, PASCAL, PL/1, APL, etc. Generally speaking for different languages of similar power, the easier it is to learn to use a language, the more successful it is taken to be.

3.4   Computers and the Chinese language

A familiar difficulty is to make a fast easy-to-use Chinese type-writer. A proposal has been made which appears to be very promising in terms of existing technology[1]. The problem faced is closely related to the matter of using Chinese language on computers. There are two somewhat different tasks. One task is the use of Chinese language rather than (say) English in the programming languages. A different task is to store and process texts in Chinese language in the computer.

The first task is more open to *ad hoc* solutions because what is needed is generally easy-to-memorize reminders and requires only a very limited vocabulary or abbreviations. It is then fairly simple to select a limited set of Chinese characters and break each in a natural way into an unambiguous ordered set of strokes. To input one of these characters we have only to type in the strokes in the given order and they will be stored in the computer in a suitably coded form just as all information has to be stored in some coded form. Since the set of characters employed is fairly small, to recover their familiar form by an output device is also a relatively easy task.

When the goal is to store and recover extensive texts in Chinese, the questions involved are about the same as making a typewriter that is fast and easy to use. Since the set of characters is large, it would be tiresome to require a fixed ordering of the strokes which make up a given character and, moreover, there is the problem that a set (ordered or not) of strokes may correspond to more than one character.

---

1)   See the paper with B. Dunham (in Chinese), *Dou Sou Bimonthly*, no. 14 (March, 1976), pp. 56—62. The English version came out as an IBM internal report *A recipe for Chinese typewriters* (RC4521) in September, 1973. The idea is simply to break up each character into its strokes in the natural manner so that there is a function from characters to sets of strokes. The keys of the typewriter are these strokes and other common symbols (numerals and punctuation marks). The computer part of the typewriter locates the character(s) consisting of the strokes and goes to the subroutine for writing it out (after a selection from the display in case of ambiguity). The crucial gain in speed over typewriters (for English say) is obtained by the possibility of simultaneously striking several keys to input several strokes much as the player of a piano would ordinarily hit several keys at each moment.

In the case of the typewriter, this question of ambiguity can be resolved with the help of a display device so that the typist can select the intended character whenever ambiguities arise. For the computer input, it would seem necessary to do a careful analysis of characters in the Chinese language and introduce suitable additional distinguishing features to eliminate ambiguities resulting from different characters consisting of the same set of strokes. It seems likely that attractive solutions to this problem can be found so that characters can be easily put into the computer without ambiguity and the display device becomes unnecessary for the typewriter.

The problem of recovering the familiar form of the characters through the output device encounters the central difficulty of getting good Chinese typewriters because one would like to avoid the cumbersome practice of providing one special physical "lead character" for each character. The solution of this difficulty for the typewriter is to use a more advanced printing device so that the codes for the strokes are sufficient to locate a small subroutine which enables, for example, the inkjet printing device to write out the desired character. Actually for the output device of the computer, if the technology of inkjet or similar printing is not available, we could afford to add a more traditional printing device with its operations controlled by computer instructions. In that case it is of course desirable to make the output device detached from the main computer since the physical motion of bringing the selected "lead character" to the desired position is necessarily much slower than operations inside the computer.

In short, the discussion here intends to argue that the Chinese language does not create serious difficulties with regard to either the development of easier to use programming languages or the use of computers to handle natural language (Chinese in this case).

3.5   Some examples of computer applications

The field of computer science is quite heterogeneous and there is no unified theory to speak of in the present stage of its development. For this reason and for the reason that only a brief discussion is envisaged, it seems appropriate just to select a few examples to illustrate the problems in this field.

One familiar area is to apply computers in the design of computers known under the name of CAD (computer aided design). A traditional example is to use computers to simplify Boolean expressions which represent computer circuits. As it comes to large scale integration, one encounters silicon chips each containing thousands of gates. As a result, the combinatorial possibilities become so large that it is often necessary to use computers and to use them ingeniously. For example, once a number of copies of a chip are made, a computer program is employed to test which of them are the acceptable ones. There is also the problem of "placement" when the logical structure of a large circuit is determined. In order to realize the logical structure physically, it is necessary to arrange the various units appropriately so that, for example, wires or lines do not cross one another too often. Once the physical requirements are specified precisely, the problem becomes a purely combinatorial one. Since, however, the possible combinations are so large in number, it is not possible to try out all possibilities even on a large computer. Rather one has to think hard to design programs which need not give the best possible solutions but would yield good enough solutions using long but not prohibitively long computer time. It happens in this area that programs are constructed which work well in practice although no satisfactory theory exists which would determine the exact range of their applicability.

A different kind of use of computers is in storing and retrieving technical and scientific information, for example, on aeronautics and astronautics. A special organization is set up to collect and digest all openly published material on the subject in the whole world. All the material is stored in a computer. Indices and digests are compiled partly with the help of computers and published semi-monthly. Those who desire detailed information can then get it directly through a terminal or indirectly ask the organization to retrieve and send it along.

A totally different kind of application of computers is to suggest theoretical problems to be studied. If we add the time element to the logical circuits, we arrive at what is called sequential circuits. This suggests a new theory which has come to be known as the monadic

to the colleges A and B, but β prefers A to B and A prefers β to α.

Clearly any assignment under which the above situation occurs is not satisfactory to A and β. It is desirable to prevent such a situation and use only stable assignments.

Definition 2. Two tables T and T' are equivalent, $T \sim T'$, if they have the same stable assignments.

Definition 3. For any table T, its canonical form T^* is the result obtained by making all possible applications of I and II in any order we wish. In other words, T^* is the closure of T under I and II.

In order to justify the definition of T^*, it is necessary to give an argument to prove that T^* is uniquely determined by T, or, in other words, that we obtain the same result no matter in what order the operations I and II are repeatedly applied. We shall leave out the argument.

Theorem 1. $T \sim T^*$.

It suffices to prove that no stable assignment is lost by applying I or II to any given table. We illustrate the argument by showing that if T' is obtained from T by an application of II, then $T \sim T'$.

It is sufficient to prove that no stable assignment assigns to A any β of the type specified under Operation II above. Suppose the contrary. Since there are on $L(A)$ $q(A)$ applicants before β who have A as the first choice, at least one of them, say α, must be assigned to another college B. But A prefers α to β and α prefers A to B (since A is α's first choice).

For each college list $L(A)$, let $f(A)$ be the number of applicants on $L(A)$ whose first choice is A, and let $d(A)$ be the number of applicants on $D(A)$.

Theorem 2. In T^*, $f(A) = d(A)$ for every college A.

Proof. Since Operation II can no longer be applied to T^*, we have, for every A, $d(A) \geq f(A)$. Let d be $\Sigma d(A)$, over all colleges, and $f = \Sigma f(A)$, over all colleges. Then $d \geq f$.

On the other hand, every applicant on some $D(A)$ has a first choice (either A or some other college B) and appears on that college's list. Moreover, since Operation I is no longer applicable, no applicant in some $D(A)$ can appear in any other $D(B)$. Therefore, $f \geq d$. Hence, $f = d$.

Suppose for some A, $f(A) \neq d(A)$. Since, for every B, $d(B) \geq$

$f(B)$, we must have $d(A) > f(A)$. But then since $f = d$, there must be some B, $D(B) < f(B)$, contradicting $d(B) \geqslant f(B)$.

Therefore, $d(A) = f(A)$ for every A.

Corollary 1.   In T^*, every remaining applicant, i.e., one who appears in at least one college list $L(B)$, must appear in some $D(A)$.

If there were a student α who appears only on waiting lists, α must have a first choice A and appear on $L(A)$, because otherwise A would no longer be a choice for α. But that would make $f > d$.

In other words, even though it is quite possible for T to contain lists such that there is some α who only appears on waiting lists, this is no longer true for T^*.

Corollary 2.   In T^*, if for a college A, $L(A) = D(A)$, i.e., the college A has only $\leqslant q(A)$ applicants remaining on $L(A)$, then all applicants on $L(A)$ have A as first choice, and exactly these applicants will be accepted by A in any stable assignment.

Since the proof of Theorem 2 does not assume that $D(A) \neq L(A)$, we have $d(A) = f(A)$ also for such short lists. Hence, every applicant on A has A as first choice.

Since Operation I is no longer applicable to T^*, these applicants do not appear on any other college lists. Since they have A as first choice and make up $D(A)$, it is impossible for A to accept any other applicant in any stable assignment.

Corollary 3.   Every stable assignment for T assigns exactly the applicants appearing in T^* to the colleges appearing in T^* in such a way that each student α is either assigned to the college A with α in $D(A)$ or to a college B which α prefers to A.

Since $T \sim T^*$, they have the same stable assignments. Suppose α is assigned to C (in particular, to no college at all) and α prefers A to C. Since α appears on $D(A)$, A must have accepted some applicant β lower than α on $L(A)$. But then α and A can make a deal to upset the assignment.

No crucial information is lost in the transition from T to T^*. But once we arrive at T^*, there are many different possible choices. The two clearest choices are: assign to each college A, the applicants on $D(A)$ in T^*; assign to each college A, the applicants on $L(A)$ with A as first choice. For obvious reasons, the first is the college optimal assignment and the second is the student optimal assignment.

If to be judicial is to be "impartial" to the students and the colleges, then one seems to need something inbetween. But then it is no longer clear how one can find a completely satisfactory unique solution.

Various other results have been obtained but we shall only mention, without proof, a few of them for illustration.

Theorem 3. All and only stable assignments in T^* are obtained by any sequence of applications of the following two operations:

(i) Delete a terminal segment from each student's list and close under Operation I (use the college optimal assignment for the concluding step).

(ii) Delete a terminal segment from each college list and close under Operation II (use the student optimal assignment for the concluding step).

In fact, each stable assignment in T^*, can be obtained by a single application of (i); and similarly with (ii).

A terminal segment of a student's list is either empty or any unbroken part of the list including the last college on it but excluding the first college on it. Similarly, a terminal segment of a college list $L(A)$ is either empty or any unbroken part including the last name on it but excluding $D(A)$. This theorem supplies a general method of generating stable assignments which include the college optimal and the student optimal as the two extreme cases.

Theorem 4. The college optimal assignment is optimal for the colleges and "anti-optimal" for the students; the student optimal assignment is optimal for the students and anti-optimal for the colleges.

Thus in no stable assignment can the college do better than in the college optimal one because $T \sim T^*$ and each college A gets its top choices in T^*. Similarly the students can do no better than using the student optimal assignment since they all get their first choices in T^* and $T \sim T^*$.

Theorem 5. If a appears in T^*, then listing more choices initially below his last choice in T cannot help; but it can hurt under the college optimal assignment, and it cannot hurt under the student optimal assignment.

Theorem 6. Adding unattainable choices by a student can help and cannot hurt.

The difficulty in always finding solutions satisfactory to both students and colleges becomes more obvious when we consider the simpler special case known as the marriage problem.

Assume given m men $\alpha, \beta, \ldots$ and k women $A, B, \ldots$ each with a nonempty ranking list of some members of the opposite sex. Delete from every list $L(a)$ everybody whose list does not include a. Clearly anybody not appearing on anyone's list can just as well have his or her list deleted. Let the given collection of lists be T.

We now apply Operations I and II to get T^*. The situation is more symmetrical than the matching of students with colleges, so that the two operations are merged into one. Let a be a male or a female, if a is the first choice of somebody b on $L(a)$, delete all names below b on $L(a)$ and delete a from their lists. This includes I (for males) and II (for females). T^* is obtained after all possible applications of I and II are made in any order.

In T^*, everybody a who remains (i.e., occurs on any list at all) appears as first choice on exactly one list and has his or her list $L(a)$ containing exactly one name (the last one) whose first choice is a. In particular, unless a person becomes the first choice of somebody after all possible applications of I and II, he or she is eliminated. It follows also that in T^* the remaining men and women are equal in number.

It is immediately clear that situations can arise where we have no unique best matching. Take, for example, two couples making opposite choices: A prefers α to β, B prefers β to α; but α prefers B to A, β prefers A to B. In fact, this adverse pattern could occur with any number of couples.

3.7   Proof of the four-color theorem[1]

The four-color problem was first proposed in a letter in 1852 by Francis Guthrie, a mathematics student, and eventually publicized by Arthur Cayley at the London Mathematical Society in 1878. In June 1976 Wolfgang Haken and Kenneth Appel succeeded, by es-

---

1)   This section is drawn from Kenneth Appel's expository paper, *New scientist*, 21 October, 1976, vol. 72, pp. 154—155.

sential use of computers, in proving the conjecture. The place where they work, Urbana in Illinois, is so proud of the achievement that we find "four colors suffice" stamped on outgoing mail, together with the usual post marks.

As is generally familiar, the problem is to determine whether four colors are sufficient to color any map on a plane so that no adjacent regions have the same color. It is easily seen that at least four colors are needed. For example, it is easy to draw a map with four regions so that each is adjacent to all the others.

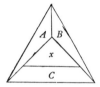

Even though a lot of work has been done toward the problem over about one-hundred years, the final solution could be viewed as an elaborate patching up of a flaw in one of the earliest attempts to prove the theorem. In 1879 A. B. Kempe published an ingenious "proof" which contained a major flaw exposed later by P. J. Heawood in 1890. Kempe argued as follows. If there are maps requiring five colors, there must be a smallest such map, that is to say, one with the smallest number of regions which requires five colors (called a minimal five-chromatic map). But no such map can exist because given any such map, we can always "reduce" it, i. e., find a smaller one also requiring five colors.

Kempe broke up the above argument into four lemmas. (1) Every map contains a region with five or fewer neighbors. (2) No minimal five-chromatic map can have a region with just two or just three neighbors (because we can then find a smaller map also requiring five colors). (3) Likewise no such map call have a region with exactly four neighbors. (4) No such map can have a region with exactly five neighbors. It turns out that Kempe did prove (1), (2), (3) but his proof of (4) is fallacious. The form of Kempe's reasoning reveals two components: (1) gives an "unavoidable" set of configurations; (2), (3), (4) purport to show that each configuration in the above set is "reducible". In general, any unavoidable set of reducible configurations would prove the theorem by showing the nonexistence of any minimal five-chromatic map. The solution obtained in 1976 is in fact the exhibition of such a set.

The task was, therefore, to study what sets are unavoidable and what configurations are reducible with the view of finding a suitable collection of configurations such that each of them is reducible and all of them taken together form an unavoidable set. Heinrich Heesch initiated the use of computers on this problem in a serious manner and found certain easily checkable properties which determine many configurations which are likely to be reducible. Haken began to search for unavoidable sets with the help of the advanced knowledge of reducible configurations, and then Appel joined him in 1972. In the process sophisticated computer programs were developed to carry out the procedure of finding candidates for an unavoidable set of reducible configurations.

The sophistication is needed firstly to enable the generation of different candidates by minor variations in input and certain parameters, and secondly to replace unwanted configurations (unwanted because not reducible or not easily seen to be reducible) without having to discard a whole promising candidate in which many configurations are already known to be reducible. For example, after a configuration is generated, if it cannot be shown to be reducible within 30 minutes on an IBM 370/168 computer, it is replaced by other configurations. In June 1976 after using over 1000 hours of computer time, an unavoidable set of under 2000 reducible configurations was produced.

The achievement of obtaining their particular solution clearly gives an impressive example of the interaction of mathematicians with computers. Computer programs were repeatedly modified as the understanding of the task increased with closer examination of the conceptual problems and the computer outputs. The use of computers is crucial and interesting because the computer time needed is large yet not prohibitive. Generally the crucial consideration in finding new uses of computers is precisely to find and define tasks which are too tedious (or, better, as in this case, practically impossible) for people yet do not require prohibitively long computer time.   It is well known that many theoretical algorithms require exponential time and are way beyond the capacity of computers. On the other hand, we also see, especially in the area of "artificial intelligence", silly experiments with computer programs which lead nowhere, merely because not enough preparatory thought has been given to the problem under consideration.

3.8   Computer proof of theorems

Within the wide range of computer applications, the attempt at assisting the proof of mathematical theorems has a history of about twenty years.   We have just discussed the striking recent success of computers as an auxiliary tool in proving the four color theorem. This type of local auxiliary use of computers as an aid to proving theorems had been made from time to time, notably by D. H. Lehmer[1]. It takes the form of singling out appropriate parts which call for extended numerical or combinatorial computations to supplement the conceptual flow of the arguments leading to a proof of a theorem. To describe such uses, we may speak of opportunist or *ad hoc* mechanization of theorem-proving.

A different direction may be called systematic mechanization which devises a general method for dealing with a domain of related problems such that, though the method need not solve every problem in the domain, it solves each problem completely whenever applicable. Ideally we should like to see a lot of interactions between these two directions.   At any rate, for the purpose of studying systematic mechanization, it seems generally desirable to experiment with moderately difficult specific examples.   A general method may be a decision procedure for a given class of problems so that by applying the method each problem of the class can be decided in the sense that we eventually settle whether it is a theorem.   The method may also be one which can only yield a definite answer for some problems in the class.   In both cases, how fast the method can be carried out is of crucial importance.

---

1) See, for example, references in his paper "Some high-speed logic", *Proc. symposia in applied math.*, vol. 15 (1963), pp. 141—145.

For example, a well known refinement of Sturm's theorem yields the decidability of elementary geometry and elementary algebra (i.e., the theory of real closed fields). Yet it is known that the theory in question can have no practicable decision procedure because any general method must take at least 2^{cn} (c a positive constant) to decide problems of length n (i.e., n is the number of symbols in the sentence expressing the problem)[1]. By the way, several people[2] have found decision procedures for the theory which take at most $2^{2^{kn}}$ (k constant) steps to decide every sentence of length n. The length of a sentence is calculated in terms of the primitive notation so that it is generally much longer than the usual form which commonly uses many abbreviations (through definitions). When exponential time is necessary, even large computers are useless since 2^j increases very fast as j increases.

Recently Wu Wen-tsun has obtained a fast (by hand) decision procedure for an interesting subdomain of the above theory, namely, elementary geometry with the "betweeness" relation disregarded[3]. Surprisingly the area includes difficult theorems such as Simson's theorem (the feet of the perpendiculars to the sides of a triangle from a point on its circumscribed circle are collinear) and Feuerbach's theorem (The nine point circle of a triangle is tangent to the inscribed circle and to each of the three escribed circles.) This result seems to be the first case in which a natural domain containing hard theorems gets a fast decision procedure. The method uses algebraic representations of geometrical problems and appears to be realizable at least partly on computers with a moderately complex program, especially if available programs for many algebraic manipulations are taken over as subroutines. It would be interesting to determine exactly how much of the method can be realized on existing computers.

---

1) See Albert R. Meyer, *Proc. Int. Cong. Math.* 74, p. 419 where the result is attributed to M. J. Fisher.

2) See, for example, G. E. Collins, "Quantifier elimination for real closed fields", *Proc. EUROSAM 74, ACM SIGSAM Bull.*, vol. 8 (1974), pp. 80—90.

3) Wu Wen-tsun, "The decision problem of elementary geometry and mechanical proof" (in Chinese), *Chinese science*, November 1977, no. 6, pp. 507—516.

An approach to systematic mechanization with special attention to mathematical logic and inferences was initiated earlier on. In the summer of 1958, I wrote three programs dealing with the first order logic; some loose ends were tied up in the autumn of 1959. These results were published in two papers at the beginning of 1960, generating considerable interest[1]. The programs were written in a rather primitive language (*SAP*, the share assembly programming language) and on a small, slow machine (the *IBM* 704). As is well known, the first order logic is undecidable. Yet all the theorems (over 350) of the first order logic in *Principia mathematica* were proved in less than nine minutes[2]. A surprising realization is that all the theorems of the first order logic actually proved in *PM* are of a rather trivial sort because they all belong to a simple decidable subdomain. This phenomenon may be viewed as an easier analogue of Wu's treatment of an efficiently decidable subdomain of elementary geometry and points to the desirability of selecting appropriate subdomains or partial methods for a theory which is either undecidable or not practicably decidable.

Since 1960, there have been a number of studies concerned with proving theorems on computers. Unfortunately a majority of published papers seem to center around the first order logic and primarily to elaborate a general theorem of Herbrand's on the first order logic. In my opinion, no striking results have appeared along this direction which is different from what I envisaged as the promising approach. I believed and continue to believe that in order to prove theorems in a special mathematical discipline, one must pay

---

1) ''Toward mechanical mathematics'' (written in the autumn of 1958), *IBM Journal*, vol. 4 (1960), pp. 2—22; ''Proving theorems by pattern recognition I'', *Communications ACM*, vol. 3 (1960), pp. 220—234. For example, several papers in the same issue as the second paper and elsewhere are devoted to introducing new program languages and they all include programs for an algorithm introduced in the first paper to show how powerful their languages are.

2) It is reported that when Russell heard about an account of some far less satisfactory result on proving theorems of PM by computers, he wrote: ''I wish Whitehead and I had known of this possibility before we both wasted ten years doing it by hand. I am quite willing to believe that everything in deductive logic can be done by a machine.'' (Ronald W. Clark, *The life of Bertrand Russell*, 1976, p. 548.)

special attention to what is specific to the subject. For example, in dealing with number theory one must pay special attention to mathematical induction perhaps in the form of trying to derive a contradiction from the assumption that there is a least counterexample to the general proposition $\forall xPx$ to be derived. There is also the idea of finding a method to obtain a formally rigorous proof from a sketch of a proof. Some of these opinions were explained at a meeting in 1965 by considering a few examples which were thought to be close to what could be handled on computers[1]. My impression is that these examples remain open problems for proofs on computers.

1) ''Formalization and automatic theorem proving'', *Proc. IFIP Cong. 1965*, vol. 1, pp. 51—58.

4. PROBLEMS AND SOLUTIONS

4.1 Problems as a driving force

In the natural sciences, theory can be applied to explain natural phenomena or to solve practical problems such as making useful artifacts, preventing and curing diseases, increasing production, etc. Both theoretical and practical problems serve to focus the efforts by trying to make use of available facts to achieve a fixed goal. This general situation also holds for mathematics except that we often encounter in mathematics interesting problems which are more remote from phenomena in the natural world and from practical applications.

It is not very informative to say, for example, that in studying number theory we are trying to find out consequences of the Peano axioms or that in studying analysis we are trying to understand the concept of real numbers or that one of the main aims of number theory is to study the prime numbers. We get a clearer idea of a subject when we know a number of its typical theorems and problems. Both in studying known results and in making new discoveries, keeping in mind a few sharp problems and thinking of their solution as the goal to be sought for serve to supply an organizing principle and a direction for learning known facts and finding new facts.

In studying mathematics, one is generally faced with a mass of known facts and a mass of open problems. Typically one makes some selection and then concentrates one's attention on one or a few problems. When one or several interrelated appropriate problems are chosen, an effort is made to absorb known facts which are relevant to these problems. There is also a period of ''incubation'' when the material is to be digested. As the attack on given problems is pursued, new problems are usually generated. The original problems either get solved or get reformulated or get cast aside.

Scientific research is a purposeful activity; problems, serving as purposes in view, play a central role in the advance of mathematics and of the work of an individual investigator. There are a number of famous problems which are widely known and hard to solve, seeing that they become famous because they have remained unsolved after repeated strenuous efforts. The selection of a particular problem to study requires preliminary examinations as well as judgments as to the available data and ability.

Two of the best known famous problems are Fermat's and Goldbach's conjectures which can be stated very easily. The former says that for $n > 2$, there are no positive integers x, y, z such that

$$x^n + y^n = z^n.$$

The latter says that every even integer greater than 2 is the sum of two prime numbers (this is often expressed briefly as "$1 + 1$"; the best result so far is $1 + 2$ by Chen Ching-yun, viz. the sum of a prime and a product of at most two primes).

Another famous problem which is of more central importance to mathematics is Riemann's hypothesis. This says that all nontrivial zeros of the function

$$\zeta(s) = \sum_{n=1}^{\infty} n^{-s}$$

of the complex variable s lie on the line with $\mathrm{Re}(s)$ (the real part of s) $= 1/2$. This is equivalent to saying that $\zeta(s)$ never vanishes in the stripe $1/2 < \mathrm{Re}(s) < 1$.

Even though these problems remain unsettled today, they have led to a great deal of interesting developments in mathematics. For example, in trying to settle Fermat's conjecture, Kummer was led to the introduction and development of ideal numbers of which some of the laws have been generalized to any algebraic field by Dedekind and Kronecker. Important contributions to mathematics have been made by proving analogues of the Riemann hypothesis (for example, for varieties over finite fields).

A common feature of the three conjectures is their relative simple logical form. They are all of the form $\forall n F(n)$ in the language of Peano arithmetic such that for each fixed integer k, $F(k)$ can be decided by elementary calculations. This is not so obvious for the Riemann hypothesis which, however, is known to be equivalent to the following statement[1]:

RH. For all n, $\left(\sum_{k \leqslant \delta(n)} 1/k - n^2/2 \right)^2 < 36n^3$.

where $\delta(x) = \prod_{n < x} \prod_{j \leqslant n} \eta(j)$, $\eta(j) = 1$ if j is not a prime power,
and $\eta(p^k) = p$ for every prime p. For brevity, we refer to the
statement as:

$$\forall n R(n).$$

For the Fermat conjecture, it should be noted that there are standard
ways to contract x, y, z, n to one variable so that $\forall x \cdots \forall u F(x, \cdots, u)$ can be treated as of the form $\forall n F(n)$.

A consequence of this common form is the following remark.
If any one of these conjectures is undecidable in the formal system
PA (or even suitable small fragments of it) described in the second
chapter above, then it is true. The developments so far have not
yielded methods which appear applicable in showing any of the three
conjectures undecidable in PA. But if a proof of undecidability
(say of RH) is found, then we get also a proof of the proposition
itself. The reason is quite simple, as we briefly indicated before.
For example, if RH is not true, then there is some k, such that
$R(k)$ is not true. But then since $R(k)$ is decidable by simple calculations, $\neg R(k)$ is a theorem of PA by a well-known general fact
about PA. Hence, $\exists n \neg R(n)$ or $\neg \forall n R(n)$ would be a theorem
of PA. Therefore, if RH is undecidable in PA, then RH is true.

The most famous list of open problems is perhaps the one given
by Hilbert in 1900. A symposium was held in May, 1974 with a
large number of mathematicians to consider the mathematical consequences of the Hilbert problems. The reports include not only
references to solutions and partial solutions of the problems obtained
to date but also mainly "an attempt to focus upon those areas of
importance in contemporary mathematical research which can be seen
as descended in some way from the ideas and tendencies put forward
by Hilbert in his speech." The proceedings of the symposium[2] con-

---

1) See the paper of Davis-Matijasevicz-Robinson in Browder 1976, the book
listed in footnote 2) on page 11.
2) Browder 1976.

tains both a separate list of problems of present-day mathematics and also open problems in the various reports each dealing with matter related to one (group) of Hilbert's original problems. It is clear from the volume how great a role the Hilbert problems have played since 1900. On the other hand, "we should not fail to note that there is no hint in them of such decisive developments in the following decades as the development of topology, both combinatorial and set-theoretic, or of functional analysis."

In Hilbert's list of twenty-three problems and groups of problems, four are directly related to mathematical logic. The first problem is the continuum problem which led to the central results in axiomatic set theory and continues to dominate work in the area. It remains unsolved in its original form as first proposed by Cantor in 1878. The second problem is the consistency of analysis which is open to diverse interpretations and has led to a large body of developments under the heading of proof theory. The tenth problem on the solution of Diophantine equations has been settled with the emergence of the theory of algorithms (or recursion theory). The seventeenth problem on the sums of squares[1] has stimulated some of the more interesting work in model theory. These problems are touched on elsewhere in these lectures in the appropriate contexts.

4.2   Problems in mathematical logic

Within each discipline there are large problems which may or may not take on a sharp form but the solution or partial solution of them would affect broad areas in the discipline. For example, the continuum problem and the attempt to find structures which resemble the constructible sets in crucial aspects but allow larger cardinals would seem central to set theory. There is the desire to further

---

1) In this connection, it is interesting to note that an important contribution to the problem was made by the following paper by C. Tsen.

[27]   C. Tsen, "zur Stufentheorie der Quasi-algebraisch-Abgeschlossenheit kommutativer Körper", *J. Chinese Math. Soc.* vol. 1 (1936), pp. 81—92.

To quote, "the first important step after Artin was taken by Tsen [27] in 1936 with his very general result on quasialgebraically closed fields.... Apparently his paper had been forgotten during the war, so that his results had to be rediscovered (in 1952)" (see Browder 1976, p. 485 and p. 487).

develop model theory so that striking particular constructions of models such as those giving strong independence results find their natural place in a general theory. The wish to find ways to prove familiar mathematical propositions undecidable in formal systems of number theory or beyond has stimulated many attempts ever since 1931. An area which has been around for some time but has only recently received wide attention has to do with the complexity of computations and proofs. A striking particular problem is to determine whether there is a fast general method for deciding whether a formula of the propositional calculus (or a Boolean expression) is a tautology. It is not known either that there is a method doable in polynomial time or that there is no method performable within quadratic time.

There are also a large number of special problems which one encounters as a more detailed study in a discipline is undertaken. Of course it serves little purpose to present such long lists in a series of popular lectures. It suffices to mention a few existing lists in a footnote[1].

When the statement of a mathematical problem can be understood by the nonspecialist, there is a special fascination. For example, the four-color conjecture which has been proved recently with the help of computers has a wide appeal. Of course in this as in most cases, the proof is not easy to understand. Apart from the contrast between statement and proof, generally in mathematics there are both hard proofs using simple machinery (notably many combinatorial problems) and easy proofs using complex machinery (much of algebraic topology being possibly an example). As a subject de-

---

[1] I have run across four fairly long lists:

(1) A general list covering different areas of mathematical logic by Harvey Friedman, ''One hundred and two problems in mathematical logic'', *J. symbolic logic*, vol. 40 (1975), pp. 113—129.

(2) A list of 24 open problems in model theory, Chang-Keisler, *Model theory*, 1973, pp. 512—514.

(3) A list of 43 open problems in set theory by A. R. D. Mathias, ''The real line and the universe'', *Logic colloguium* 76, ed. R. Gandy and M. Hyland, 1977, pp. 536—543.

(4) A list of 50 open problems in *Word problems,* ed. W. W. Boone, F. B. Cannonito, R. C. Lyndon, 1973.

velops, more machinery tends to be added so that the apprentice period tends to get longer. Of course, within a given subject there is also a choice between problems which require more machinery and those which require less.

Mathematical logic is a relatively young subject. For many years one did not need extensive knowledge to do interesting work in it. This has gradually changed over the years. Between 1920 and 1940, Skolem was able to introduce several simple basic ideas more or less from scratch[1]. While Gödel's works between 1930 and 1940 called for unusual originality and ingenuity, they did not require knowing a wide range of mathematical facts. As more and more people got into the field over the past twenty years or so, some aspects of mathematical logic have become more like traditional areas of mathematics in requiring extended preliminary training. However, with the possible exception of set theory, the different branches of logic are probably still more directly accessible. This seems true for the study of the complexity of computations and proofs. With model theory the basic results remain limited and mostly quite transparent especially for those who have a moderately good background in algebra.

Even with set theory, it remains true that the statement of many central problems and results is relatively easy to understand. Moreover, the basic concepts and theorems deal with a subject matter which is vaguely familiar to nonspecialists. For example, the continuum problem asks just how many real numbers or points on the line there are. It requires some explanation before one can understand a more exact statement of this problem, which says, briefly, every infinite set of reals is either countable or as large as the set of all reals. But it is a long way from understanding the statement to getting the proof. We do not know how long the journey will be since we have not reached the proof yet. Another example is the measurable cardinals. Roughly speaking, we envisage a cardinal whose subsets can be divided into large and small ones in such a way that putting countably many small sets together we still get a small set. Much work has been done on such cardinals and we do not real-

---

1) Compare the survey paper in Skolem's *Selected works in logic*, 1970.

ly know whether they exist. Yet another example is the axiom of determinacy which has been studied extensively with complex results but which is easy to state. Take a set ‘A of real numbers between 0 and 1. Players I and II play in turn 0 or 1 each time until an infinite sequence of 0's and 1's is obtained. Player I wins if the real number corresponding to the sequence belongs to the original set A. Otherwise II wins. The axiom says that for every game (i.e., every set A), either I or II has a winning strategy, i.e., a function from finite strings of 0 and 1 to the set consisting of 0 and 1 so that he will always win by playing according to it.

4.3 Some relatively transparent problems

Rather than discussing here these central concerns of today's set theory, I shall list and comment on a few computational or algorithmic or combinatorial results and problems which can be stated in a more transparent manner.

4.3.1 *Post's special tag problem*[1].

One type of problem is to ask whether certain sentences follow from the axioms in a formal system and to find a general method to decide for every sentence expressible in it whether it is a theorem. This type of problem is illustrated in an idealized manner by a simple but nontrivial game-like problem first put forth by Post in 1921.

We consider finite strings of 0's and 1's. The idea is to cut off the head of a given string (the first three symbols in the following example) and add different suitable symbols at the end of the string depending on whether the original string begins with 0 or 1. Let the tag system consist of the following two rules:

$$0\_\_ \rightarrow 0\ 0$$

$$1\_\_ \rightarrow 1\ 1\ 0\ 1$$

In other words, if a string begins with 0, cut off the first three symbols and add 0 0 at the end; otherwise add 1 1 0 1 at the end.

---

1) See Davis 1965, *The undecidable*, p. 372. For the unsolvability of certain classes of tag problems, see the last section of Appendix C.

If the string has less than three symbols, stop. Here we have an obvious concept of "deduction": if we can arrive at B from A by repeated applications of the two rules, then B is deducible from A. One open problem is: is there a general method to decide for any strings A and B whether B is deducible from A? Another is: do we have a general method to decide for each string A, whether it has infinitely many different "consequences"?

4.3.2   *Thue sequences.*

Consider the question of finding an infinite string made of a fixed finite set of symbols such that no consecutive part is of the form EE, i.e., no finite part is immediately repeated. It is easy to see that we cannot find such an infinite string using only two symbols. Thue was able to do this with three symbols. We shall outline some of the results along this line here and leave more detailed considerations in section 2 of Appendix A.

Consider first strings made out of 0 and 1. Let $R_1(0, 1) = 0$, $R_{n+1}(0, 1) = R_n(01, 10)$. It is easy to verify that for $n < m$, R_n is an initial segment of R_m. Hence, we can define an infinite string as the union of all the finite strings R_1, R_2, etc. If we represent this string as $e_0 e_1 e_2 \ldots$, we can see that it is defined by the following property: for every n, $e_n = 0$ or 1 according as the sum of the digits of the binary notation of n is even or odd. This sequence R has the property that there is no continuous stretch EEE or even EEe such that e is 0 or 1 and E begins with e.

A simpler construction than Thue's original one yields the following sequence T in three symbols which contains no EE.

(2a)   Let $T = d_0 d_1 d_2 \cdots$, where $d_n = a$ if $e_n e_{n+1} = 01$, $d_n = b$ if $e_n e_{n+1} = 10$, $d_n = c$ if $e_n e_{n+1} = 00$ or 11 in the sequence R. Thus, $T = acbabcabacb \cdots$; there is no EE in this sequence.

One application of the sequence R is a way of covering the plane with black and white pieces such that no rectangular array with 4 or more pieces forms a "torus". (See Appendix A.)

This example illustrates in an idealized form the manipulation with symbols which is prominent in computer programming and in some parts of mathematical logic. The application to covering the plane is related to certain questions in finding finite models and periodic solutions for sentences in the predicate calculus.

One of the achievements of mathematical logic is the development of tools to prove the nonexistence of algorithms (the unsolvability) of certain classes of problems. These problems are often easy to state although they are not easy to settle.

4.3.3 *Word problem for groups and Burnside's problem*[1].

The word problem for groups was first proposed by Max Dehn in 1911. It asks whether for every finitely presented group G there is a general procedure by which we can decide whether two words are equal. Around 1955, this was answered in the negative by giving a group G whose word problem is unsolvable. A finitely presented group is a group given by finitely many generators, $x_1, \ldots, x_n$ and finitely many relations. More explicitly, a word is nothing but a finite string obtained by using the n generators as basic symbols. The relations stipulate that certain combinations of words are identical (e.g., $AB = BA$ for all words when the group is Abelian). Being a group implies that there is a unit (an identity element) and that every element has an inverse. These restrictions are absent for semigroups. It is this difference that accounts for the fact that it is much easier to show that the word problem for semigroups is unsolvable. It was proved in 1947 by a more or less straightforward representation by a semigroup of a Turing machine which generates a recursively enumerable set that is not recursive[2].

In 1902, Burnside proposed the question: Is every group finite when it has a finite number m of generators and satisfies the identity relation $x^n = 1$?

This was answered negatively in 1968: for arbitrary $m > 1$ and arbitrary odd $n \geqslant 4381$, there is an infinite group satisfying $x^n = 1$.

By the way, the Burnside problem has little to do with mathematical logic except that much of the same kind of consideration is employed as in the study of the word problem.

In an earlier attempt to settle the Burnside problem it was claimed that for $m \geqslant 3$, $n \geqslant 72$, there are suitable groups G of exponent n with m generators such that:

---

1) For extensive expositions see *Word problems* listed in footnote on page 53.

2) See, e.g., Davis 1965, p. 292.

(a)   Every Dyck word W (i.e., $W =$ unit) in G contains a block of the form EE.

Hence, if we apply T of (2a), we obtain the desired results because, e.g., all initial segments of T must be distinct by (a). Hence, applying (2a), the Burnside problem reduces to proving (a).

4.3.4   *The string Diophantine problem.*

We shall say more about the Diophantine problem (Hilbert's tenth problem). First we mention an open problem which is an analogue of it for the "concatenation" operation in place of addition and multiplication.

Consider now the problem with strings obtained from the two symbols a and b. For example,

$$axb = bxa$$

has no solution, while

$$axbay = yaxab$$

has solutions, e.g., $x = ba$, $y = ab$. Now we can ask whether an arbitrary finite set of string equations (or, equivalently, a single equation) has solutions. There are several interesting partial results. The general problem is open.

For example, it is easy to show that the general problem reduces to the case of finite sets of equations in which each equation contains no more than three variables. On the other hand, a fairly complex argument shows that any finite set of equations each containing no more than two variables is decidable[1].

4.4   The Diophantine problem

Example. $3x = 2y + 1$

$$x = \frac{2y + 1}{3} \quad y = 1, 4, 7, 10, \cdots$$

$$x = 1, 3, 5, 7, \cdots$$

----

1)   A book centering around this problem was published in 1971 and translated into English in 1976: Ju. I. Hmelevskii, *Equations in free groups*, iii + 270 pp. (Am. Math. Soc., Providence, R. I.).

Hilbert's tenth problem is equivalent to asking whether there is a general method by which, given any equation E:

$$P(x_1, \cdots, x_n) = 0$$

where P is an arbitrary polynomial with integers coefficients, we can decide whether E has a solution in nonnegative integers. Up until about 1935, no clear meaning was available for a negative solution to the problem because we had no precise concept of "general methods". In 1970 it was proved that there exists no general method for solving Diophantine equations[1].

A useful concept is obtained by dividing the variables in a Diophantine equation into two sets $\{a_1, \cdots, a_m\}$ and $\{u_1, \cdots, u_n\}$. For brevity, we shall take $m = 1$:

$$(1) \qquad P(a, u_1, \cdots, u_n) = 0.$$

We shall say that a set S of natural numbers is Diophantine if there is a Diophantine equation (1) such that $a \epsilon S$ iff (1) has a solution in the unknowns $u_1, \cdots, u_n$ or equivalently:

$$(2) \qquad \exists u_1 \cdots \exists u_n (P(a, u_1, \cdots, u_n) = 0).$$

In our consideration of formal systems, we have hinted at the fact that the property of being a proof in a formal system is decidable (or recursive) and that the property of being a theorem is listable (or recursively enumerable, briefly r.e.). The second part follows from the fact that every theorem is the last line of some proof and that we can list or effectively enumerate all proofs of a formal system.

The basic result on Diophantine equations is this:

4.4.1   Every r.e. set is Diophantine and there is a procedure such that for every r.e. set S of natural numbers, there is an equation (1) for which (2) is true iff $a \in S$.

Since a basic result of the theory of algorithms is the existence of r.e. sets which are not decidable, an immediate consequence is that

1)   For a good exposition of the solution, see Martin Davis, "Hilbert's tenth problem is unsolvable", *Amer. math. monthly,* vol. 80 (1973), pp. 233—269. For details and references relating to what is sketched in the rest of 4.4, see the article mentioned under footnote 1) on page 51.

there is some particular P such that the problem of deciding whether any a satisfies (2) is unsolvable. Hence, there is not even an algorithm for testing for solvability of a particular one parameter family of Diophantine equations, not to say one for all Diophantine equations.

A particularly surprising result is:

4.4.2   The set of primes is Diophantine.

The proof of this is a little involved and there are attempts to find polynomials with the least degrees and the least number of unknowns. For example, several "simple" polynomials determining the set of primes have been found: (a) one with 12 unknowns, (b) one with 21 unknowns and of degree 21, (c) one with 325 (occurrences) of symbols, etc[1].

Typically when one has shown a type of problem to be unsolvable, there is the question of finding the boundary between the decidable and the undecidable. For example, how small a universal Turing machine can one find (measured by the number of states and the number of symbols)? How small can an undecidable subset of the sentences of the predicate calculus be? More often than not, such questions do not have clean answers unless one has selected certain natural class and ask simply whether that class is large enough.

With regard to Diophantine equations there are interesting results obtained quite independently of mathematical logic:

4.4.3   Every Diophantine equation can be reduced to an equation of degree $\leqslant 4$.

4.4.4   There is an algorithm for Diophantine equations of degree 2 in arbitrarily many unknowns.

While 4.4.3 is a brief observation of Skolem, 4.4.4 is a substantial mathematical result[2]. A natural open question is:

4.4.5   *Open problem.*

---

1)   See Browder 1976, p. 331. In the same exposition (on p. 347) the existence of a Diophantine equation of the form (1) is given such that (2) is true iff there is a map which cannot be colored with a colors. Hence, when $a = 4$, the truth of the four color conjecture implies that the equation has no solution when $a = 4$.

2)   For 4.4.3, see p. 351, op. cit. 4.4.4 is proved by Carl Ludwig Siegel, "Zur Theorie der quadratischen Formen", *Nachr. Akad. Wiss. Göttingen Math.-Phys. Kl. II* (1972), pp. 21—46.

Is Hilbert's tenth problem decidable for equations of degree 3?

Another easily stated open problem is:

4.4.6   Is there an algorithm for deciding whether an arbitrary Diophantine equation has a solution in rational numbers?

This is known to be equivalent to the decision problem for non-trivial solutions in integers of homogeneous Diophantine equations.

So far we are concerned only with theoretical algorithms, paying no attention to how long it takes to decide each question. There are many situations where we know we can eventually decide each problem in a given class but we are interested in finding out whether there is a fast or at least feasible method to do it. I conclude with a pair of examples illustrating this type of problem.

4.5   Euler paths and Hamilton paths

In the eighteenth century, seven bridges in a Köngisberg park suggested a general problem: to decide whether a given graph has an Euler path. i.e., a way of beginning from one node and going along all lines so as to travel through each line exactly once.

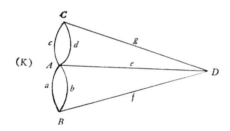

Of course, there is a theoretical method for deciding this class of problems: given any (finite) graph, try all the (finitely many) ways of travelling through the graph from a fixed node as origin; either we find an Euler path and stop to give the answer "yes", or we have tried all paths but found no Euler path and stop to give the answer "no". When the graph has many lines, this is too tedious because of the large number of combinations.

Euler found a better way. He discovered two simple conditions which are necessary and sufficient for a graph to contain an Euler path: it must be connected, i.e., we can get to any node from any other node; and every node, except for possibly two, must be at the junction of an even number of lines. For example, the graph (K), though connected, fails the second condition.

Using these conditions, we get a much more efficient method for deciding whether a graph contains an Euler path.

Hamilton asked a related question on graphs: does a given graph have a Hamilton path, i.e., a path which passes through each node exactly once (but not necessarily traversing each line of the graph)? For example, the graph (K) does have a Hamilton path: e.g., go from C to A to B to D. No fast general method is presently available for deciding whether any graph contains a Hamilton path. All the known algorithms for this class of problems use an exponential amount of time so that if the number of nodes is n, there is no polynomial $P(n)$ in n which, for all n, would give an upper bound to the time required to answer the question for every graph with n nodes.

There are many types of problem which are in the same situation as the one about Hamilton path. One example is to decide whether a Boolean expression in conjunctive normal form has a model.

5. FIRST ORDER LOGIC

5.1 Satisfiability and validity

The predicate calculus is known under various different names: first order logic, predicate logic, elementary logic, the restricted functional calculus, the restricted predicate calculus, the theory of quantification with identity, relational calculus, etc. First order logic seems to be the fashionable usage among logicians today.

If we wish to study a collection of relations (or properties which may be viewed as special, namely unary, relations), we generally think of a universe of discourse A and consider which n tuple from A (i.e., which x in A^n) satisfies R. In terms of linguistic expressions, we ask when an (atomic) sentence $R(a_1, \ldots, a_n)$, for $a_1, \ldots, a_n$ in A, is true. But we are also interested in more general statements such as there is a smallest positive integer or there is no largest integer, or every integer is either even or odd, etc. Moreover, we are interested in making inferences to assert that a certain sentence is true if certain other sentences are true. In fact, the matter of inference is regarded as the central concern of logic.

When we think of mathematical discourses, we encounter all the time neutral particles such as the "quantifiers" $\forall$ (all) and $\exists$ (some), the connectives $\wedge$ (and), $\vee$ (or), $\neg$ (not), $\supset$ (only if), and the symbol $=$ for identity as well as variants such as for all, there exists, iff, etc. It is easily granted that these occur in scientific discourses of the most diverse sorts and can all be taken as logical constants. It is harder to convince oneself that these (or just some of them, say $\exists$, $\supset$, $\neg$, $=$) are sufficient to yield all the constants of logic. Only recently has there been work to show that under reasonable general conditions, these are all[1].

---

1) Compare the general discussion in *From mathematics to philosophy*, pp. 143—163. See also section 5.8 below in this chapter.

When we develop a scientific theory, there are typically a domain of objects we intend to study and a collection of statements to be asserted or denied. In the simplest case we would assume that every object has a name and ask whether certain relations hold for certain objects. It is natural to introduce the propositional connectives to say, for example, that A is older than both B and C, B is not older than A, etc. Typically, we would like to make general statements about a collection of objects which may be infinite and it is natural to use quantifiers to do this. The identity symbol is useful because we are often interested in knowing whether different naming expressions, say $2 + 2$ and $3 + 1$, name the same object. Functions such as $+$ enrich the domain of naming expressions. For example, the first order system PA for the natural numbers (see Chapter 2) intends unsuccessfully to capture all the true statements about natural numbers.

We can also make an abstraction to write down "schemata" in which relation and function symbols (as well as individual constants) are not given interpretations in advance but rather ask to be interpreted:

(1) $\forall x \, \exists u \, \forall y \, [\, \neg \, Rxx \land Rxu \land (Ruy \supset Rxy)]$,

(2) $\forall x \, \exists u \, \forall y \, [(Ryx \supset \neg \, Rxy) \land (Rxy \lor Ryu)]$,

(3) $\forall x \, \exists u \, \forall y \, [(Ruy \supset \neg \, Ryu) \land (Rxy \lor Ryu)]$,

(4) $\forall x \, \forall y \, \forall z \, \exists u \, [f(x, 0) = x \land f(x, u) = 0 \land f(x, f(y, z))$
$$= f(f(x, y), z)].$$

With some reflection it can be verified that each of (1), (2), and (3) is satisfied when we take an infinite domain as the range of variables and an ordering relation $<$ as R. In particular, we may choose ω as the domain and write briefly $\langle \omega, < \rangle \models$ (1), to express that (1) is satisfied by the structure $\langle \omega, < \rangle$, or alternatively, that the latter structure is a model of (1). The statement (4) is the condition for groups so that any structure that is a group satisfies (4); for example, take all integers with zero for 0, $+$ for f. By the way, since (1), (2), (3) cannot be satisfied in a finite domain, they are sometimes called axioms of infinity.

It is fairly easy to work intuitively with the notions of satisfiability and model. A related notion is validity. If $\neg A$ has no model, then A is said to be valid; or, if A and B have the same models, $A \equiv B$ is valid. A great deal can be said about logical (or quantificational) schemata without reference to any (complete) formal system of first order logic; the latter becomes crucial when we wish to tie up validity with provability and satisfiability with consistency.

As a first example, we consider the prenex normal form, i.e., the form in which all the quantifiers are at the beginning of the sentence. We take $\neg$, $\vee$, $\wedge$, $\forall$, $\exists$, $=$ as the only logical constants and assert that each schema W built up with them can be turned into a schema W′ in prenex form such that W and W′ have the same models. For this purpose, we observe that the following schemata are valid: $\neg\forall x A \equiv \exists x \neg A$, $\neg\exists x A \equiv \forall x \neg A$, $\neg(A \wedge B) \equiv (\neg A \vee \neg B)$, $\neg(A \vee B) \equiv (\neg A \wedge \neg B)$, $(\exists x A \vee B) \equiv \exists x (A \vee B)$, $(\exists x A \wedge B) \equiv \exists x (A \wedge B)$, $(\forall x A \wedge B) \equiv \forall x (A \wedge B)$, $(\forall x A \vee B) \equiv \forall x (A \vee B)$. It should be clear that a quantifier in any context can be moved forward by these rules until it is at the beginning or only preceded by quantifiers.

This simple result has wide applications in the classification of schemata and predicates (such as the arithmetical hierarchy) because we can concentrate on the quantifiers as a measure of complexity. Generally speaking, a class K of logical schemata is called a reduction class with respect to satisfiability (resp. validity) if for every schema W, there is a corresponding $W′$ in K such that W is satisfiable (resp. valid) iff $W′$ is. Thus the class of schemata in prenex form is a reduction class with respect to both satisfiability and validity. Since about 1920, there has been a good deal of work on reduction classes.

5.2 Reduction classes and the decision problem of first order logic

The decision problem of first order logic has several equivalent formulations in terms of satisfiability, validity, provability or refutability. We consider just one formulation. The problem is to find a general method by which, given any logical schema, one can decide whether it is satisfiable. This general problem is known to be unsolvable. For example, in Gödel's famous paper of 1931, Theorem X says that we can find, for each (primitive) recursive predicate F, a corresponding logical schema W such that $\forall xFx$ is true iff W is satisfiable. Hence, if there were a decision procedure for logic, the set of x such that $\forall y \neg T(x, x, y)$, with the predicate T of Appendix C, would be a recursive set, contradicting the result that $\{x \mid \exists y T(x, x, y)\}$ is not recursive. In 1936, Turing showed the undecidability by reducing the halting problem to it with a device similar to the representation of tiling problems by logical schemata in Appendix C.

There remains the problem of deciding subclasses of the class of logical schemata. The most familiar classification is by using the class of formulas in the prenex form. Relative to this classification, the questions are by now almost completely answered in the sense that every prenex class is either decidable or a reduction class (and hence undecidable). The steps in this result are relatively few and can be briefly summarized.

5.2.1   The class of schemata of the form

(1) $\exists x_1 \cdots \exists x_m \forall y_1 \cdots \forall y_n U(x_1, \cdots, x_m; y_1, \cdots, y_n)$

with identity included is decidable for satisfiability. In fact every satisfiable schema of the form (1) has a finite model[1].

5.2.2   The class of schemata of the form

(2) $\exists x_1 \cdots \exists x_m \forall y_1 \forall y_2 \exists z_1 \cdots \exists z_n U(x_1, \cdots, x_m, y_1, y_2, z_1, \cdots, z_n)$

not containing identity is decidable for satisfiability. In fact, every satisfiable schema of the form (2) has a finite model[2].

---

1)   The case without identity was proved in P. Bernays and M. Schön-finkel, *Math. Annalen*, vol. 99 (1928), pp. 401—419. The case with identity was established in F. P. Ramsey, *Proc. London Math. Soc.*, vol. 30 (1929), pp. 338—384. The latter paper includes the introduction and proof of the famous Ramsey Theorem to be considered later.

2)   This was first proved by Gödel, *Monatsh. Math. Phys.*, vol. 40 (1933), pp. 433—443. The result had been presented to Karl Menger's Kolloquium in 1930.

The following open problems remain: (a) Whether the class of schemata of the form (2) remains decidable when identity is included; (b) Whether a satisfiable schema (2) with identity always has a finite model. Various unsuccessful attempts have been made to settle these problems.

On the side of prenex form reduction classes, a surprising result was obtained in 1961 which unites all such reduction classes. There is no need to include identity[1].

5.2.3   The schemata of the form

$$(3) \qquad \forall x \exists u \forall y U(x, u, y),$$

with dyadic predicates only (or even a single dyadic predicate plus monadic predicates) make up a reduction class for satisfiability.

Given this, the most interesting previously known reduction class follows fairly directly. We give the derivation of the following corollary to illustrate a type of classical arguments.

5.2.4   The schemata of the form

$$(4) \qquad \forall x \forall u \forall y \exists w U(x, u, y, w)$$

make up a reduction class for satisfiability.

It is sufficient to find a schema (4) for each schema (3) so that (4) is satisfiable iff (3) is. Take a new predicate A and consider:

$$(5) \qquad \forall x \exists w A x w \wedge \forall x \forall u (A x u \supset \forall y U(x, u, y)).$$

If (4) is satisfiable, then (5) is, by interpreting Axu as $\forall y U(x, u, y)$. If (5) is satisfiable, then (4) is because we can even derive (i.e., infer by transformations preserving all models) (4) from the two halves of (5). But (5) can easily be rewritten as $\forall x \forall u \forall y \exists w (Axw \wedge (Axu \supset U(x, u, y)))$, which is of the form (4). Hence, type (3) is reduced to type (4).

---

1)   The sort of consideration leading to the proof of this theorem is illustrated in Appendix C. For full proofs, see *Proc. Nat. Acad. Sci.*, vol. 48 (1962), pp. 365—377 and *Math. Theory of Automata*, Polytechnic Institute of Brooklyn, 1963, pp. 23—70.

Combining 5.2.1—5.2.4, we get a general theorem, if we disregard identity:

5.2.5 Given any string of quantifiers $Q_1x_1\ldots Q_kx_k$, the prefix class determined by it is a reduction class relative to satisfiability if and only if the string contains $\forall\exists\forall$ or $\forall\forall\forall\exists$ as an order-preserving but not necessarily unbroken substring. Each prefix class that is not a reduction class is decidable and contains no axioms of infinity (i.e., no schemata which are satisfiable but not satisfiable in finite domains).

Thus, by 5.2.3 and its corollary 5.2.4, if the string contains $\forall\exists\forall$ or $\forall\forall\forall\exists$, then it is a reduction class. Otherwise, it either contains three or more consecutive $\forall$'s but followed by no $\exists$'s, or contains at most two consecutive $\forall$'s but no other $\forall$'s. In the first case, it falls under the general case 5.2.1. In the second case, it falls under the general case 5.2.2. In both cases, the absence of axioms of infinity is also explicitly established.

We have concentrated on decidability for satisfiability. Since a schema W is not satisfiable iff $\neg W$ is valid, the results can be transformed simply into ones about validity. Moreover, once a formal system for first order logic is set up and shown to be complete, it follows immediately that semantic validity equals syntactic provability, semantic satisfiability equals syntactic consistency (see 5.3.4 below). Probably the most interesting decidable subdomain of first order logic not under the prenex classification is the monadic first order logic[1].

Much more can be done without appeal to a formal system, for example, the two proofs of the Löwenheim-Skolem theorem. Since, however, it is convenient to tie up these proofs with completeness and later concepts introduced in model theory, we shall first

---

[1] There is a large literature on these decision problems and reduction problems. On *Solvable cases of the decision problem*, there is W. Ackermann's book of Amsterdam, 1954; and currently Burton S. Dreben and Warren D. Goldfarb are preparing a more systematic book. On reduction problems, results before 1960 are put together in Janos Suranyi, *Reduktionstheorie des Enischeidungsproblems*, Budapest, 1959; a more up-to-date book is under preparation by Harry R. Lewis. An extensive consideration of monadic first order logic is contained in Ackermann's book.

say something about the formalization of the propositional calculus which does not require much machinery.

5.3 The propositional logic

There is a wide range of studies in propositional logic[1] which is, however, no longer in the main stream except for the new interest in finding faster methods to decide all propositional schema. We can only make a quick survey of some of the many different aspects.

There are different sets of adequate primitive connectives. For example, if we use the notation $|$ for "not both" and $\downarrow$ for neithernor, each of the following sets has been used: $\{|\}, \{\downarrow\}, \{\neg, \vee\}$, $\{\neg, \wedge\}, \{\supset, \neg\}, \{\neg, \vee, \wedge\}, \{\neg, \vee, \wedge, \supset\}, \{\neg, \vee, \wedge, \supset, \equiv\}$. Once a set of primitive connectives is chosen, there are different formal systems. For each formal system, there are the questions of consistency, independence, and completeness.

For illustration, let us consider a familiar system based on $\neg$ and $\vee$. We have a sequence of propositional letters and generally a schema is defined inductively: every propositional letter is a schema, if U and V are schemata, so are $\neg U$ and $U \vee V$.

5.3.1 *System P_1.*

Rule of inference: if U and $\neg U \vee V$, then V.

Axioms.   (1) $\neg (U \vee U) \vee U$;

              (2) $\neg U \vee (U \vee V)$;

              (3) $\neg (U \vee V) \vee (V \vee U)$;

              (4) $\neg (\neg U \vee V) \vee (\neg (W \vee U) \vee (W \vee V))$.

The intended interpretation of the system is given by the familiar method of "truth tables" which also gives a method of deciding whether a schema is valid. We have just two truth values t (true) and f (false). $v(\neg U) = t$ iff $v(U) = f$, $v(\neg U) = f$ iff $v(U) = t$; $v(U \vee V) = f$ iff $v(U) = v(V) = f$, $v(U \vee V) = t$ otherwise. A schema is valid or a tautology iff it gets the value t, no

---

1) To give just one example, a book of almost 600 pages is entirely devoted to the propositional logic: H. Arnold Schmidt, *Mathematische Gesetze der Logik: I. Vorlesungen über Aussagenlogik*, 1960.

matter what values (t or f) each propositional letter in it may take. It is easy to verify that all axioms of P_1 are valid. Moreover, if $\neg U \vee V$ and U are valid, then V is valid. Hence, all theorems of P_1 are valid.

It follows immediately that P_1 is consistent both in the sense that not all schemata are theorems and in the sense that for no schema U, are both U and $\neg U$ theorems. The axioms can also be shown to be independent of one another by using more truth values than two (in fact, three values happen to suffice for these axioms)[1]. Moreover, it can also be shown to be complete in the sense that all tautologies are theorems. Instead of proving the completeness of P_1, we prove it for a different system both to illustrate another approach and to give a more direct argument.

The alternative approach uses sequents of schemata and has the property that no schema is ever deleted in the process of a derivation. We take the schemata as in P_1. Each schema U is a string; if A, B are two strings then A, U, B is a string and $A \rightarrow B$ is a sequent. In particular, we permit the trivial case of an empty string. The definition of validity of a sequent $A \rightarrow B$ is: $v(A \rightarrow B) = t$ iff either some schema in A is false or some schema in B is true, $v(A \rightarrow B) = f$ otherwise. The system P_2 has five rules altogether: one initial rule, two rules for $\neg$, and two rules for $\vee$.

5.3.2   *System P_2.*

P_2 1.   Initial rule.   If A, B are strings of proposition letters (atomic formulas), $A \rightarrow B$ iff some proposition letter occurs on both sides.

P_2 2.   If $A, B \rightarrow C, U$, then $A, \neg U, B \rightarrow C$; if $U, C \rightarrow A, B$, then $C \rightarrow A, \neg U, B$.

$P_2$3.   If $A, U, B \rightarrow C$ and $A, V, B \rightarrow C$, then $A, U \vee V, B \rightarrow C$; if $C \rightarrow A, U, V, B$, then $C \rightarrow A, U \vee V, B$.

This system has a simple decision procedure (for provability) which yields also a proof of completeness. Thus given any sequent,

---

we find the first connective $\neg$ or $\lor$ (if any) and apply $P_2 2$ or $P_2 3$ in reverse order to eliminate it. In this way, we obtain one or two sequents each with one less occurrence of a connective. When the process is repeated, we finally obtain a finite set of sequents in which no more connectives occur. A sequent is a theorem iff all the resulting sequents are theorems according to $P_2 1$. It is easy to see that if any of the resulting sequents is not a theorem according to $P_2 1$, we can so choose the truth values of the proposition letters that it gets the value f and, therefore, the conjunction of all these simple sequents gets the value f. And then by using these truth values, the original sequent also gets the value f. Hence, all valid sequents are provable in P_2, and P_2 is complete.

5.3.3 The system P_2 is complete.

Propositional systems of this particular kind were first introduced in connection with proving theorems of logic on computers[1] and they are well liked by computer scientists because they are algorithmically transparent.

We make explicit the relation between consistency and satisfiability mentioned in the previous section:

5.3.4 If a formal system S of (propositional) logic is complete, then a schema is consistent iff it is satisfiable (or, every schema is either satisfiable or refutable in S).

We take for granted the fairly obvious result that in each of the systems we consider, only valid schemata (or sequents) are derivable. Hence, if a schema W is satisfiable, then $\neg W$ is not provable because it is not valid. Hence, adding W yields no contradiction and, therefore, W is consistent. Conversely, since the system is complete, if W is not satisfiable, then $\neg W$ is a theorem. Hence, adding W gives a contradiction and W is not consistent.

Traditionally a generalization of the argument for completeness

---

1) The general form of such systems for first order logic (including quantifiers) goes back to Gentzen's paper of 1935; see The collected papers of Gerhard Gentzen, 1969. A special simplicity is attained when propositional logic is separated out with appropriate revisions to avoid rules which would infer shorter sequents from longer ones. This simplification was first introduced in *IBM journal of res. and dev.*, vol. 4 (1960), pp. 5—8.

yields a generalization of 5.3.4 to the effect that a (countably many) infinite set of schemata is either satisfiable or contains a refutable finite subset. Such a generalization follows immediately from the following theorem[1].

5.3.5   *Compactness for satisfiability.*

A set T of schemata is satisfiable iff every finite subset of T is.

For the simple case of propositional logic, we can think of an enumeration of all schemata in T and consider the sequence $W_n = U_1 \wedge \cdots \wedge U_n$. If every finite subset of T is satisfiable, then, for each n, W_n is satisfiable. We can arrange all satisfying truth assignments for W_1, W_2, etc. in an infinite tree form. Then the hypothesis of the infinity lemma (see Appendix A) is satisfied and we have, therefore, some infinite path which gives a model for the whole set T. Since there are familiar ways of replacing a schema with quantifiers by a countable set of propositional schemata, a similar argument can also prove 5.3.5 for schemata with quantifiers.

A more popular proof is to extend T to a maximal set whose finite subsets are all satisfiable. A set S is maximal iff, for every schema U, either $U \in S$ or $\neg U \in S$. It is easily seen that a maximal set S with all finite subsets satisfiable is satisfiable by the simple device of assigning t to every member of S. The assignment satisfies the basic requirements because $\neg$ is satisfied by definition and $v(U \vee V) = t$ iff $v(U) = t$ or $v(V) = t$ by a simple argument: Suppose otherwise, we would have $\{U \vee V, \neg U, \neg V\} \subseteq S$ or $\{\neg(U \vee V), U, \neg V\} \subseteq S$ or etc. But then we would have a finite subset of S which is not satisfiable.

To obtain a maximal extension S of T, assume all schemata are enumerated: W_1, W_2, etc. Let $T = S_0$ and $S_{n+1} = S_n \cup \{W_{n+1}\}$ if every finite subset of this is satisfiable and $= S_n \cup \{\neg W_{n+1}\}$ otherwise. In the latter case, every finite subset of $S_n \cup \{\neg W_{n+1}\}$ must be satisfiable because, by definition, $S_n \cup \{W_{n+1}\}$ has a finite subset P which is not satisfiable. If there were also a finite subset Q of $S_n \cup \{\neg W_{n+1}\}$ not satisfiable, then $R = P \cup Q - \{W_{n+1}\} - \{\neg W_{n+1}\}$, a subset of S_n, would also not be satisfiable because any

---

1) See Theorems IX and X in Gödel's 1930 paper, *Monatsh. Math. Phys.*, vol. 37, pp. 349—360.

satisfying assignment to propositional letters in R (or an extension thereof) must make either W_{n+1} true or $\neg W_{n+1}$ true.

Let $S =$ the union of all S_n and we have the desired maximal set. It is possible to extend also this argument to the full first order logic[1]. We shall make a few general remarks about model theory before considering such an extension and its applications.

5.4 Model theory

In model theory one studies the interaction of language and interpretation in the special case of models of theories formalized in the framework of formal logic, especially that of first order logic. A first order language is given by a set of symbols for relations, functions, and constants which in combination with the basic symbols of first order logic determine in a familiar manner certain combinations of symbols as sentences (formulas, well formed formulas). Speaking abstractly, a theory is but a collection of sentences and a structure is a set of structured objects. A structure S is a model of a theory T if S makes all members of T true. We review some of the central parts of model theory in the remainder of this chapter, even though the considerations in the earlier sections of this chapter can also be said to belong to model theory.

It is better to begin with an example. Consider the theory of ordered groups.

5.4.1 The theory T of ordered groups consists of the set of all theorems derivable by first order logic from the following axioms:

1) Group axioms. $\forall x \forall y \forall z [x + (y + z) = (x + y) + z \wedge x + 0 = x \wedge \exists u(x + u = 0)]$.

2) Order axioms. $\forall x \forall y \forall z [((x < y \wedge y < z) \supset x < z) \wedge (x \neq y \equiv (x < y \vee y < x))]$.

3) $\forall x \forall y \forall z [(x < y) \supset (x + z < y + z \wedge z + x < z + y)]$.

For this theory, any structure of the form $S^* = \langle S, <, +, 0 \rangle$, i.e., a set S in which one object is singled out as 0, a relation on S^2 is taken as $<$, a function from S^2 into S is taken as $+$, is a candidate for being a model for T. If S^* satisfies all sentences in T, then

---

1) Such an extension goes back to Leon Henkin, *J. symbolic logic*, vol. 14 (1949), pp. 159—166.

it is a model of T expressed by $S^* \models T$. (Following common practice, we shall often disregard the distinction between S and S^*.) For example, the set of all integers with the familiar 0, $<$, and $+$ is a model of the above theory T.

With good reasons it is sometimes said that model theory is a branch of set theory and proof theory is a branch of number theory. In considering structures and models, we take for granted our intuitive understanding of sets. Hence, given a structure associated with a fixed vocabulary, the set of true sentences is already determined and in such a way that the requirements from first order logic are satisfied (e.g., $\neg A$ is true iff A is false, etc.). For example, the true sentences of arithmetic in the vocabulary of PA (see Chapter 2) constitute a theory which is not recursively axiomatizable.

5.4.2   *Theories.*

A theory T is complete if it consists of exactly the true sentences of some structure S. It is finitely axiomatizable if it is equivalent to a finite set of sentences (i.e., it consists of exactly theorems provable by first order logic from a finite set of axioms). It is recursively axiomatizable if it is equivalent to a recursive set of sentences (in a recursive language).

There are many examples: the theories of groups, rings, fields, ordered fields, Abelian groups, lattices, and Boolean algebras all are finitely axiomatizable; the theories of divisible groups, algebraically closed fields, finite fields, real closed fields, Peano arithmetic and ZF set theory are all recursively axiomatizable but not finitely axiomatizable; the theories of finite groups, arithmetic (i.e., all true sentences rather than just theorems of Peano arithmetic) and set theory (i.e., all true sentences of ZF rather than just theorems of ZF) are not recursively axiomatizable.

Take for example the theory T of divisible groups. The axioms are those in 1) of 5.4.1 plus an infinite set of axioms:

(1) $\forall x \exists y (y + y = x)$, $\forall x \exists y (y + y + y = x)$, etc.

Suppose there were a finite set of axioms. They could also be written as a single axiom A which is true of all divisible groups. Hence A is provable in T and there is a finite set F of axioms T

such that $F \vdash A$. Only finitely many members of (1) can occur in F. Suppose it includes only those with no more than k y's. Take a prime $p > k$. The cyclic group of order k satisfies F and A but is not divisible. Hence, T. is not finitely axiomatizable.

We have mentioned the compactness theorem. There are a few other conceptually simple theorems and methods of construction which are central to model theory. Examples are the Löwenheim-Skolem submodel theorem, Skolem functions, ultraproducts, diagrams, Ramsey's theorem and homogeneous sets, model completeness, saturated models, omitting types theorems, etc. We shall consider only a small part of model theory and refer the reader to more extensive treatments in the literature[1]. It should be mentioned that much of model theory is related to algebra. It is sometimes said that model theory is the union of logic with universal algebra.

Let us return to a more general consideration of the compactness theorem 5.3.5. Traditionally one is accustomed to using only countably many symbols in a language (in particular, in a first order language) specified in a recursive way. But it has been common practice in model theory to imagine also uncountably many symbols (e.g., constants) and prove more general theorems. We shall primarily be concerned with countable languages and theories but we can illustrate the extension to uncountable languages in the case of the compactness theorem.

Let L be, for example, the first order language of set theory (i.e., with the dyadic relation $\in$ as the only additional symbol beyond the basic symbols of logic) and M be a structure for this language L.

5.4.3 The diagram language of M is the expansion L_M of L formed by adding a new constant c_m for each $m \in M$. The diagram of M is the complete theory determined by M in L_M, i.e., the set of all true sentences in the language of L_M satisfied by M.

Using this natural concept, it is easier to state a proof of the

---

1) We mention four books. C. C. Chang and H. J. Keisler, *Model Theory*, 1973. M. D. Morley (editor), *Studies in model theory*, 1973. Part A (especially the Chapter A2) of Jon Barwise (editor), *Handbook of mathematical logic*, 1977. Jane Bridge, *Beginning Model Theory*, Oxford, 1977.

compactness theorem (also known as the finiteness theorem) in its general form. (Please disregard the sketch of proof following 5.4.4 if you find it presupposing unfamiliar notions.)

5.4.4   *The compactness theorem.*

Let T be a first order theory (i.e., a set of sentences in a first order language L). If T is finitely satisfiable (i.e., every finite subset of T is satisfiable), then T has a model.

Let L have α sentences (α an infinite cardinal number) and C be a set of α new constants. Consider the expanded language L_C with α sentences well-ordered in some manner as $\{F_\beta | \beta < \alpha\}$. Take T as T_0 and define $T_{\beta+1}$ from T_β by requiring: (i) $F_\beta \in T_{\beta+1}$ if $T_\beta \cup \{F_\beta\}$ is finitely satisfiable, (ii) if $F_\beta \in T_{\beta+1}$ and F_α is of the form $\exists x G(x)$, then $G(c) \in T_{\beta+1}$ for some $c \in C$. Take union at limit ordinals. Consider now the equivalence relation for every pair c and d such that $c = d \in T_\alpha$. Take a subset M of C such that exactly one element of C is taken from each equivalence class, and T' be the subset of T_α consisting of all sentences in the language L_M. It can be shown that T' is the diagram of a model of T. The essential use of finite satisfiability is much the same as in the outline of the proof of 5.3.5 for the simpler propositional case. The addition of new constants and the condition (ii) above is to take care of the quantifiers.

The sketch permits uncountable theories. It is probably confusing for most readers. In any case, we shall primarily be interested in the countable case and a proof of 5.4.4 in that case is a simple generalization of the completeness theorem.

5.4.5   The first order Peano arithmetic PA has some nonstandard model, i.e., has some model in which there are some "unnatural" numbers.

Add a new constant c and consider the infinite sequence of sentences: $c \neq 0$, $c \neq 1$, $c \neq 2$, etc. Clearly every finite subset is satisfiable in the standard model of PA since we only have to take as c a natural number greater than all the numbers occurring in the finite set. Therefore, by the compactness theorem, the theory obtained from PA by adding all these sentences as new axioms has a model. But then there must be some object c distinct from all the usual natural numbers.

5.4.6 *Nonstandard analysis (or non-Archimedean ordered fields).* Take the structure of all the real numbers and its theory T. Add a new constant c and the axioms $c > 0$, $c < 1$, $c < \frac{1}{2}$, etc. This new theory has a model, and c is an infinitesimal or, in other words, the model must be a non-Archimedean ordered field. Recall that a field is Archemedean if for any two positive elements a and b, there exists some n, $na \geqslant b$.

5.4.7 If a theory has arbitrarily large finite models, then it has infinite model.

5.4.8 Every partial order on a set can be extended to a simple order on it.

This uses the known result that 5.4.8 is true for finite sets. Compactness yields the extension to infinite sets.

5.5 The Löwenheim-Skolem theorem

Generally we are interested in different models for the same theory so that we commonly consider structures associated with the same vocabulary.

5.5.1 A structure M is an elementary submodel of N, or $M \prec N$, iff $M \subseteq N$ and for every sentence A and objects $a_1, \ldots, a_n$ in M, $M \models A(a_1, \ldots, a_n)$ iff $N \models A(a_1, \ldots, a_n)$. N *is* also said to be an elementary extension of M.

This concept becomes clearer as we discuss one of the earliest results in model theory:

5.5.2 *The LS submodel theorem.* If a logical schema W has a model N, then it has a countable model M such that $M \prec N$. The same is true if instead of W a finite or countable set of schemata (a theory) is assumed to have a model.

Skolem begins by considering a simple case:

$$(1) \qquad \forall x \exists y U(x, y).$$

Assume that (1) is satisfied in a given domain N for certain interpretations of the predicate letters, etc. "Then, *by virtue of the axiom of choice*, we can imagine that for every x a uniquely determined y is chosen in such a way that $U(x, y)$ comes out true". This defines a mapping y_x of the domain N into itself. Then

(2) $$\forall x U(x, y_x)$$

is true for the given interpretation of the predicate letters. Let a be a particular individual and M be the intersection of all classes X such that $a \in X$ and $y_x \in X$ if $x \in X$. Then M is (either finite or) countable. But (2), and therefore (1), also hold true in M.

The function y_x or $y(x)$ in the illustration is an example of Skolem functions. The above outline of proof has many applications, refinements, and extensions. We mention some.

It follows from 5.5.2 that no countable first order theory can determine the real numbers in a unique manner since the intended model is uncountable. Similarly, if the ordinary axiom system ZF has a model, then it has also a countable model M. For example, we can prove in ZF a theorem (1) $\exists x(x$ is uncountable), i.e., ZF $\vdash$ (1) and therefore $M \models$ (1). But (1) is only true within M, from outside the model, (1) is not true for M. In other words, even though M contains many sets (e.g., ω_1 in M) which are not countable by the functions available in M, they are nonetheless countable in the "real world." This phenomenon is sometimes called the Skolem paradox. In current studies of set theory, one often encounters the contrast between the cardinality of a set in a given model and its true or real cardinality.

We mention an early application of 5.5.2 to the question of finite axiomatizability.[1]

5.5.3 Any first order formal system which is rich enough to carry out the construction of the countable model in 5.5.2 for each single sentence is not finitely axiomatizable.

Suppose such a system S had a finite set of axioms. Take F as the conjunction of these axioms, turned into the prenex form. Apply the construction illustrated in the proof of 5.5.2. By assumption, we would arrive at an enumeration of a countable model in S. But then we would have a consistency proof of S in S, contradicting Gödel's second theorem (see Chapter 2). For example, the usual

---

1) This was first pointed out in *Proc. Nat. Acad. Sci.,* vol. 36 (1950), pp. 479—484. For an elaboration of this observation as applied to ZF, see P. J. Cohen, *Set theory and the continuum hypothesis,* 1966, pp. 82—83.

system ZF of set theory is shown to be not finitely axiomatizable by this remark.

Let $|A|$ be the cardinality of a set A.

5.5.4 If N is an infinite model of a theory T, $S \subseteq N$, and $|S| \leqslant \alpha \leqslant |N|$, then there is a model M of T such that $M \prec_e N$, $S \subseteq M$, and $|M| = \alpha$.

Take any set A, $S \subseteq A \subseteq N$, $|A| = \alpha$. Take all the Skolem functions of the sentences in T and extend A to get a closure with respect to them (the Skolem hull of A). In other words, let $A_0 = A$, A_{n+1} be the union of A_n with all values of Skolem functions with arguments in A_n, and let $M = A_\omega$. Since N is closed with respect to Skolem functions, $M \subseteq N$ and $M \prec_e N$. Since countably many copies of an infinite cardinal α add up to α again, $|M| = \alpha$.

5.5.5 Let M be an infinite model of T and $\alpha \geqslant |M|$. There is an elementary extension N of M such that $|N| = \alpha$.

Add a set C of α new constants and the axioms $c \neq d$ for any distinct constants c and d in C. By the compactness theorem, the extended theory has a model.

5.5.6 If T has no finite models and T is α-categorical for some infinite cardinal α (i.e., any two models of cardinality α are isomorphic), then T is complete.

If T is not complete, then there is a sentence F such that neither F nor $\neg F$ is in T. We can therefore extend T in two different ways and, by 5.5.5, obtain two nonisomorphic extensions of cardinality α.

For example, the theories of atomless Boolean algebra and dense linear order without endpoints are ω-categorical and hence complete.

We can summarize 5.5.4 and 5.5.5 by saying:

5.5.7 If a theory T has any infinite model, then, for any infinite cardinal, T has a model of that cardinality.

It follows that no theory with any infinite model can be categorical since two models of different cardinalities can certainly not be isomorphic. A natural question is whether a theory can be categorical in certain infinite cardinalities and not categorical in others. A theorem due to Morley[1] simplifies the problem considerably.

---

1) M. Morley, ''Categoricity in power,'' *Trans. Am. Math. Soc.*, vol. 114 (1965), pp. 514—538.

5.5.8 Let T be a theory (in a countable language). If T is categorical in some uncountable cardinal, then T is categorical in every uncountable cardinal.

Hence, there are only two kinds of α-categoricity, ω-categoricity and ω_1-categoricity. Examples are known for all four combinations: there are theories that are categorical (1) in every infinite cardinality; (2) for ω but not for ω_1; (3) for ω_1 but not for ω; (4) in no infinite cardinality.

A well-known open problem is to determine in general how many non-isomorphic countable models a theory T_c can have. The conjecture is that if T has more than ω non-isomorphic countable models, then it has 2^ω such models. Of course the continuum hypothesis must not be used to settle this.

There are different ways of generalizing the LS theorem by considering properties other than the cardinality of the universe of a model. One way is to study "two cardinal problems". Suppose one of the relation symbols in the given language is a property U. By an (α, β)-model is meant a model of cardinality α in which the interpretation of U has cardinality β. Various results have been proved. For example, if T has an (α, β)-model, then (1) it has a (γ, β)-model, for $\beta \leqslant \gamma \leqslant \alpha$; (2) it has an (ω_1, ω)-model; (3) it has a (γ, δ)-model if $\alpha \geqslant \gamma \geqslant \delta \geqslant \beta$; (4) for each regular cardinal γ, it has a (γ^+, γ)-model, γ^+ being the next larger cardinal after γ; etc. There are also various independence results on the general question: which pairs of infinite cardinals (α, β) and (γ, δ) have the property that every model of type (α, β) for a countable language has an elementary submodel of type (γ, δ)? Two pairs having the property is known as Chang's conjecture for them. It has been demonstrated up to the present stage of development that we have here almost an inexhaustible source of possible problems to exercise one's expertise in mastering different techniques.

5.6 Ultraproducts

In 1933 Skolem introduced a method of constructing models for the theory of natural numbers to give nonstandard models. This was generalized by Los in 1955 and has been widely used since then[1].

Let T be the first order theory of number theory, i.e., the set of all true sentences in the language of PA (see Chapter 2). Skolem assumes a model M of T given (say the standard model) and constructs a new model consisting of functions rather than numbers which contribute a model of T but is not isomorphic with M (in fact, is of greater order type).

Skolem proves the following combinatorial lemma:

5.6.1 Let $f_1(x)$, $f_2(x)$, ... be an infinite sequence of functions of natural numbers. Then there exists a monotone increasing function $g(x)$ such that, for any pair f_i and f_j, one of the three following cases holds for almost all x (i.e., for all but a finite number of them):

$$f_i(g(x)) < f_j(g(x)), \; f_i(g(x)) = f_j(g(x)), \; f_i(g(x)) > f_j(g(x)).$$

Let $f_i \sim f_j$ if $f_i(g(x)) = f_j(g(x))$ for almost all x. In other words, if C is the set of all cofinite subsets of ω:

5.6.2 $f_i \sim f_j$ iff $\{t \mid f_i(g(t)) = f_j(g(t))\} \in C$.

The new model is obtained by taking the equivalence classes (by 5.6.2) of the sequence of all definable functions of one variable in T (using Skolem functions to get rid of quantifiers) so that $\omega^* = \{f_1, f_2, \ldots\}/\sim$. It is verified by using these equivalence classes, with the ordering determined by 5.6.1, that addition and multiplication can be defined over ω^* to satisfy all sentences in T. It is clear that ultraproducts are one of the several ways of generalizing Skolem's construction to yield wider applications.

We begin with ultrafilters[2].

5.6.3 *Ultrafilters.*

Let I be a nonempty set. A set D of subsets of I is an ultrafilter on I if the following conditions are satisfied:

1) I belongs to D;

---

1) See Th. Skolem, *Fund. math.*, vol. 23 (1934), pp. 150—161 and J. Los, *Fund. math.*, vol. 42 (1955), pp. 38—54.

2) For extensive treatment of the topics mentioned in the next few paragraphs, see Chapter A3 in *Handbook of mathematical logic* and J. L. Bell and A. B. Slomson, *Models and ultraproducts*, 1969.

2)   If A, B are in D, so is their intersection $A \cap B$;

3)   If A is a subset of B, and A is in D, then B is in D ;

4)   For every subset A of I, either A is in D or I minus A is in D. Roughly stated, each ultrafilter of a set I conveys a notion of large subsets of I so that any property applying to a member of D applies to I ''almost everywhere.''

This divides all subsets of I into two categories by condition 4). When one says simply that D is an ultrafilter, we imply that I is simply the union of D or the largest set in D.

Let $\{ \langle A_i, R_i \rangle \}$, $i \in I$ be a family of structures and D be an ultrafilter on I.

5.6.4   The Cartesian product B of $\{A_i\}$ is the set of all sequences f such that $f(i) \in A_i$ for all $i \in I$.

5.6.4′   For f, g in B, $f \sim g$ or f and g are D-equivalent, iff $\{i | f(i) = g(i)\} \in D$.

5.6.5   The structure $\langle W, S \rangle$ is the ultraproduct of $\{ \langle A_i, R_i \rangle \}$ iff W is the set of all f^* such that $f^* = \{g \in B | g \sim f\}$ and S is the relation such that Sfg iff $\{i | R(f(i), g(i))\} \in D$.

In the special case when $\langle A_i, R_i \rangle$ all are a same structure $\langle A, R \rangle$, the structure $\langle W, S \rangle$ thus defined is called the ultrapower of $\langle A, R \rangle$. When D is just a filter, but need not be an ultrafilter, $\langle W, S \rangle$ is called the reduced product of $\langle A_i, R_i \rangle$ or the reduced power of $\langle A, R \rangle$.

There are two fundamental theorems on these concepts.

5.6.6   If $\langle A_i, R_i \rangle$ $(i \in I)$ are structures over the same language, then for every sentence F in the language:
$\langle W, S \rangle \models F$ iff $\{i | \langle A_i, R_i \rangle \models F\} \in D$. In particular, if each $\langle A_i, R_i \rangle$ is a model of a theory T in the language, then $\langle W, S \rangle$ is also a model of T.

5.6.7   A necessary and sufficient condition for two structures over the same language to be elementarily equivalent (i.e., to have the same set of true sentences) is that they admit ultrapowers which are isomorphic.

One application of these theorems is in the introduction of non-standard analysis, which was originally instituted by other considerations. By using a suitable ultrapower of the structure of the field R of real numbers, a real closed field that is elementarily equivalent to R is obtained that is non-Archimedean. This development supplies an unexpected exact foundation for the classical differential calculus using infinitesimals, which has considerable historical, pedagogical, and philosophical interest[1].

A widely known application to algebra is to Artin's conjecture on the field Q_p of p-adic numbers where p is any prime number. The conjecture states that every homogeneous polynomial of degree d over Q_p, in which the number of variables exceeds d^2, has a nontrivial zero in Q_p. It was proved in 1965 by using ultraproducts that the conjecture is true for arbitrary d with the possible exception of a finite set of primes p (depending on d). It was subsequently verified that indeed the original conjecture is not true when extended to full generality[2].

5.7　Ramsey's theorem and indiscernibles

A familiar pigeon-hole principle says that if m things are to be distributed in n drawers and $m > n$, then some drawer must contain more than one thing. For example, we can infer by this principle that there must be at least two persons who have the same number of hairs or have the same birthday (and even the same "birthsecond"). A more general principle states that, if a set of large cardinality is partitioned into a small number of classes, some one class will have large cardinality. Those elements of the set that lie in the same class cannot be distinguished by the property defining the partition.

A related concept is that of "indiscernibles", which also has rather extensive applications in set theory. An ordered subset of the domain of a structure M of a language is a set of indiscernibles for M, if M cannot distinguish the members of the subset from one

---

1)　See Chapter A6 of *Handbook*, H. J. Keisler, *Elementary calculus*, 1976, and Martin Davis, *Applied nonstandard analysis*, 1977.

2)　See J. Ax and S. Kochen. *Am. journal math.*, vol. 87 (1965), pp. 605—648, and *Annals of math.*, vol. 83 (1966), pp. 437—456; Y. Ershov, *Sov. math. dokl.*, vol. 165 (1965), pp. 1390—1393. For the counterexample to the original conjecture, see G. Terjanian, *C. R. Acad. Sci. Paris*, vol. 262 (1966), p. 612.

another. More exactly, given any $x_1 < \ldots < x_n, y_1 < \ldots < y_n$ in
the subset, then for any sentence $F(a_1, \ldots, a_n)$ of the language
$M \models F(x_1, \ldots, x_n)$ iff $M \models F(y_1, \ldots, y_n)$.

In the study of the decision problem of first order logic, Ramsey
proved a generalization of the pigeon-hole principle[1]. Let $[X]^n$
be the set of (unordered) n-tuples from x, i.e., $\{x \subseteq X \mid |x| = n\}$.

5.7.1 *Ramsey's theorem (finite case).*

Given m, n, k, there exists some p such that for every set x with
at least p members, if we divide $[x]^n$ into k disjoint sets, $C_1, \ldots, C_k$,
then there is some set T, $T \subseteq X$, $|T| \geqslant m$, and $[T]^n \subseteq C_i$, for some
i, $1 \leqslant i \leqslant k$.

5.7.2 *Ramsey's theorem (infinite case).*

If K is infinite and $[K]^m$ is partitioned into k subsets $C_1, \ldots,$
C_k, then there is an infinite set T, $T \subseteq K$ and $[T]^m \subseteq C_i$, for some
i, $1 \leqslant i \leqslant k$. (The set T is sometimes said to be homogeneous for
the partition).

The theorem holds for all m and all k. It is sufficient to
prove it for $k = 2$ and derive it for higher values of k by induction:
Suppose $k = j + 1$. By merging C_j with C_{j+1}, the case for $k = j$
gives $[T]^m \subseteq C_i$, $1 \leqslant i < j$ or $[T]^m \subseteq C_j \cup C_{j+1}$. In the first case, we
are done. In the second case, since T is infinite, it must have, by
the case $k = 2$, an infinite subset S such that $[S]^m \subseteq C_j$ or $[S]^m \subseteq$
C_{j+1}. Hence, we need consider only the case $k = 2$.

Moreover, we can also give the idea of the proof by considering
merely the case $m = 2$, because the induction step to move from
$m = i$ to $m = i + 1$ is entirely similar to the case $m = 2$, with $[K]^i$
in place of K.

Suppose first there exist a sequence of members x_1, x_2, $\ldots$ of
K together with a sequence of subsets K_1, $K_2, \ldots$ of K such that:

$$x_1 \in K, x_1 \notin K_1, K_1 \subseteq K, \text{ and } \{x_1, y\} \in C_1, \text{ for all } y \in K_1;$$

$$x_{n+1} \in K_n, x_{n+1} \notin K_{n+1}, K_{n+1} \subseteq K_n, \text{ and } \{x_{n+1}, y\} \in C_1, \text{ for all } y \in K_{n+1}.$$

Then $[T]^2 \subseteq C_1$, where $T = \{x_1, x_2, \ldots\}$. Obviously, x_1, x_2, $\cdots$ are

---

1) See footnote 1) on page 66.

all distinct, and K_1, K_2,... all are infinite.

Suppose now no such pair of sequences exist. This means that no matter which member of K is taken as x_1, there is a number n, such that, however x_2, ..., x_n, K_1, ..., K_{n-1} are chosen to ensure that K_n is infinite, K_{n+1} is finite for any choice of x_{n+1}. Take any fixed x_1 and consider its associated K_n. Since K_n is infinite, we can find y_1, y_2, ... and J_1, J_2, ... to satisfy:

$$y_1 \in K_n, y_1 \notin J_1, J_1 \subseteq K_n, \text{ and } \{y_1, z\} \in C_2, \text{ for all } z \in J_1;$$

$$y_{m+1} \in J_m, y_{m+1} \notin J_{m+1}, J_{m+1} \subseteq J_m, \text{ and } \{y_{m+1}, z\} \in C_2, \text{ for all } z \in J_{m+1}.$$

Thus, take $J_1 = \{z | z \in K_n \wedge \{y_1, z\} \in C_2\}$. J_1 is infinite because otherwise we could have used y_1 as x_{n+1} and $\{z | z \in K_n \wedge \{y_1, z\} \in C_1\}$ as K_{n+1}. Similarly, for each m, given J_m, we can take any member of J_m as y_{m+1} and take $\{z | z \in J_m \wedge \{y_{m+1}, z\} \in C_2\}$ as J_{m+1}. J_{m+1} must be infinite because otherwise we could have taken y_{m+1} as x_{n+1} and $\{z | z \in J_m \wedge \{y_{m+1}, z\} \in C_1\}$ as K_{n+1}. Hence, $T = \{y_1, y_2, \ldots\}$ and $[T]^2 \subseteq C_2$.

Intuitively it is perhaps easier to think of K as well-ordered (by the axiom of choice) and in terms of an infinite tree. Begin with the first member a of K and make two branches by $A_1 = \{u | \{a, u\} \in C_1\}$ and $B_1 = \{u | \{a, u\} \in C_2\}$. Take the first members of A_1 and B_1 as the next nodes. Repeating this process as long as we have an infinite set along each branch, we obtain an infinite tree. There must be an infinite path with all left branches or one with all right branches. Suppose there is no infinite path with all left branches, there is a node a_n with only a right branch. Beginning with a_n, we have an infinite path with only right branches, otherwise if there were any left branch anywhere, we could have attached it directly to a_n.

Ramsey's theorem has been generalized in various ways and a more general notation has been introduced[1].

5.7.3 The partition relation $\beta \rightarrow (\alpha)_\gamma^m$, where β, α, γ are ordi-

---

1) For detailed treatment of the topics outlined in the next few paragraphs, see Frank R. Drake, *Set theory*, 1974, p. 206, pp. 217—218, and Chapter 8. Compare also Chapter A5 in *Handbook of mathematical logic*.

nals and m is a positive integer holds iff, whenever K is a set of order type β and f partitions of $[K]^r$ into γ subsets $(f:[K]^m \to \gamma)$, there is a subset T of K which is of order type $\geq \alpha$ and homogeneous for f. When $\gamma = 2$, it is common to omit γ.

In this notation, Ramsey's theorem is $\omega \to (\omega)_k^m$ for all m, $k < \omega$. It is easy to show that for no cardinal α, $\alpha \to (\omega)^\omega$. Hence, there is no point in considering the relation $\alpha \to (\beta)_\gamma^\delta$ when δ is infinite. On the other hand, there is considerable interest in considering all finite subsets of a set (and especially a cardinal) simultaneously.

5.7.4   The partition relation $\beta \to (\alpha)^{<\omega}$ holds if β is a cardinal and for every partition of $[\beta]^{<\omega}$ (i.e., all finite subsets of β) into two parts, there is a homogeneous subset of β which is of order type $\geq \alpha$.

5.7.5 β is a Ramsey cardinal iff $\beta \to (\beta)^{<\omega}$.

According to this definition, it is easy to show that ω is not a Ramsey cardinal. In fact, Ramsey cardinals are very large. The interest in Ramsey's theorem is largely related to the interest in indiscernibles. A first consequence of 5.7.2 is:

5.7.6   If S is a structure and X is a new linearly ordered set, then there is an elementary extension M of S in which X is embedded in such a way that X is a set of indiscernibles for M.

Given a structure M and a set X of indiscernibles for M, there has been developed an elegant general method of generating a theory $K(M, X)$ and a substructure $H(X)$ of M such that $H(X)$ *is a* model of $K(M, X)$. Since a Ramsey cardinal yields indiscernibles for the universe of constructible sets (compare Chapter 7 for a discussion of constructible sets), the method of using indiscernibles to generate models leads to various structural results about the universe L of constructible sets. For example, given any Ramsey cardinal, truth in L becomes definable, every set definable in L becomes countable, the ordered set of uncountable cardinals in L becomes a set of indiscernibles for L, etc.

5.8   Other logics

We have noticed in Chapter 2 that the usual formulation of

Peano axioms as a categorical theory is not a first order system but rather a second order system. There are also infinitary logics. For example, some logics permit infinite conjunctions and disjunctions. The compactness theorem does not hold for such logics[1].

Sometimes it is convenient to use different kinds of variable and we have a many-sorted logic; for example, when we deal with points, lines, and planes in geometry. It is known that we can merge the ranges of different sorts of variable and introduce a property for each sort of variable. Hence, many-sorted logics are a matter of convenience but present no serious conceptual problem[2].

Other quantifiers have been considered. There are infinitely many x, there are uncountably many x (briefly, Ux), etc. In particular, an elegant formal system has been set up and shown to be complete for the quantifier Ux[3]. But clearly such a system does not satisfy the Löwenheim-Skolem theorem.

A natural question is whether there is something unique about the first order logic. The development of model theory has led to a novel result, proved in 1969 by Lindström[4]:

5.8.1 Within a broad class of logics, first order logic is the only one closed under $\wedge$, $\neg$, $\exists$ which satisfies the compactness theorem and the Löwenheim-Skolem theorem.

We shall make no attempt to report on more results in model theory but just make a general remark before going back to the more traditional matter of the completeness of first order logic.

There is a large gap between the general theory of models and the construction of interesting particular models such as those employed in the proofs of the independence (and consistency) of special axioms and hypotheses in set theory. It is natural to look for further developments of model theory that will yield more systematic

---

1) See, for example, H. J. Keisler, *Model theory for infinitary logics*, 1971.

2) For a treatment of this topic, see *J. symbolic logic*, vol. 17 (1952), pp. 105—116.

3) H. J. Keisler, *Ann. math logic*, vol. 1 (1970), pp. 1—93.

4) Per Lindström, ''On extensions of elementary logic,'' *Theoria*, vol. 35 (1969), pp. 1—11.

methods for constructing models of axioms with interesting particular properties, especially in deciding whether certain given sentences are derivable from the axioms. Relative to the present state of knowledge, such goals appear fairly remote. The gap is not unlike that between the abstract theory of computers and the basic properties of actual computers.

5.9 Formalization and completeness

In each first order language, there are valid sentences ("true in all possible worlds"): for example, $\forall x(x = x)$, $G \supset G$, etc. The task of formalizing first order logic (or setting up the predicate calculus) is to find a formal system (or a schema of formal systems) such that all and only the valid sentences are theorems. Historically such formal systems which are essentially equivalent had been found long before the question of completeness was formulated and answered. The first system was given in 1879 but the question of completeness was formulated and put forth as an open problem only in 1928. It was answered by Gödel in the next year[1].

Let us outline a proof of the completeness theorem.

5.9.1 *The Löwenheim-Skolem approximation theorem.*

If a quantificational schema is satisfied in any domain at all, it is already satisfied in a countable domain. Similarly with a countable set of schemata.

The proof begins with a single schema in Skolem form assumed to be satisfiable:

---

1) The first formal system was given in G. Frege, *Begriffsschrift*, 1879. A more or less standard formulation was given in A. N. Whitehead and B. Russell, Principia mathematica, vol. I, 1910. Another standard formulation was given in D. Hilbert and W. Ackermann, *Grundzüge der theoretischen Logik*, first edition, 1928 in which the problem of completeness was proposed as an open problem. In 1929 Gödel studied this book and solved the problem. This was written up as his doctoral dissertation at the University of Vienna. A revised version was published soon after, *Monatsh. Math. Phys.*, vol. 37 (1930), pp. 349—360. This paper contains the device for dealing with identity and, as noted before under footnote on page 72, the extension to the compactness theorem.

(1) $\forall x_1 \cdots \forall x_m \exists y_1 \cdots \exists y_n U(x_1, \cdots, x_m; y_1, \cdots, y_n)$.

Since (1) is satisfied in a (non-empty) domain D, we can take an arbitrary element of D and denote it by a symbol 1. Then there must be certain n-tuple $y_1, \cdots, y_n$ in D such that $U(1, \cdots, 1; y_1, \ldots, y_n)$ is true (in the given model). Without loss of generality, we can denote these elements by 2, ..., $n+1$. In other words, there must be ways of assigning truth-values to atomic schemata in $U(1, \ldots, 1; 2, \ldots, n+1)$ so that the whole schema gets the value "true." We say that (1) has solutions of level 1. This does not exclude the case that some or all the objects denoted by the symbols 1, 2, ..., $n+1$ might be the same. Let

$$c_1 = n + 1, \quad c_{k+1} = c_k + n(c_k^m - 1).$$

For each k, consider the conjunction K_{k+1} of all schemata, for $0 \leqslant t \leqslant k$, $U(a_1, \ldots, a_m; b_1, \ldots, b_n)$ where $(a_1, \ldots, a_m)$ is any m-tuple (say the $(t+1)$ st in a reasonable enumeration) from $\{1, \ldots, c_k\}^m$ and $(b_1, \ldots, b_n)$ is $(nt+1, \ldots, nt+n)$. A model of K_{k+1} is said to be a solution of (1) of level $k + 1$. Since (1) is satisfiable, it has solutions of level k, for every k. Consider now the totality of all partial solutions of (1), i.e., the totality of sets of truth-value assignments which are solutions of (1) of level k, for some k.

For each k, there may be many (but always a finite number of) solutions. For any two different solutions L and L' of an arbitrary level, write $L < L'$ if and only if $R(i, j, \ldots)$ gets "false" in L and "true" in L', where $R(i, j, \ldots)$ is the first atomic schema having different truth-values in L and L'. When $p < q$, every solution of level q must be an extension of some solution of level P. Hence, at each level k, there must be solutions which have infinitely many extensions (i.e., solutions of higher levels which extend them). Let these be $L_1^{(k)}, \cdots, L_r^{(k)}$ arranged in the order $<$ just introduced. For each k, $L_1^{(k+1)}$ must be an extension of $L_1^{(k)}$. Otherwise, since $L_1^{(k)}$ is indefinitely extendable, there must be $L_i^{(k+1)}$ which is an extension of $L_1^{(k)}$ such that $L_1^{(k+1)} < L_i^{(k+1)}$. Since $L_1^{(k+1)}$ is not an extension of $L_1^{(k)}$, $L_1^{(k+1)}$ must differ from $L_i^{(k+1)}$ in the assignments to the part K_k. But then there must be $L_j^{(k)}$ which is a part of $L_1^{(k+1)}$ such that $L_j^{(k)} < L_1^{(k)}$. This is not possible. Hence, $L_1^{(1)}, L_1^{(2)},$

$L_1^{(3)}$, $\cdots$ yield a simultaneous solution of (1), and, therefore, (1) is satisfiable in the domain of positive integers.

5.9.2   *A refutation procedure P.*

*P*1.   Usual rules to turn a schema into the prenex form.

*P*2.   Turn the schema into the functional form: e.g.,

$$\forall x \, \forall y \, \exists u \, \forall v \, \exists w \, U(x, y, u, v, w) \quad \text{into}$$

(2) $U(x, y, f(x, y), v, g(x, y, v))$.

*P*3.   Use $0, f(0,0), g(0,0,0), \ldots$ to perform the role of $1, 2, 3, \ldots$ in the preceding proof and define solutions of level k, for each k.

*P*4.   Try successively for each k, whether (2) has a solution of level k.

A schema is refutable in P if and only if there is some k such that its prenex form has, for some k, no solution of level k.

5.9.3   The refutation procedure P is complete: i.e., a schema W is refutable by P if and only if it is not satisfiable.

The proof of this theorem is direct from 5.9.1. Thus, if W is satisfiable, we see from the previous arguments that W has a solution of level k, for every k, and, therefore, that W is not refutable by P. On the other hand, if W is not refutable by P, then it has a solution of level k, for every k. Hence, by the previous arguments, W is satisfiable (in fact, in a countable domain).

This differs from the completeness theorem only in the use of an unfamiliar refutation system in place of a familiar proof system. In fact, it is a relatively small step to establish the completeness of a proof system S (say any one of those mentioned in footnote on page 88) by showing that if $\neg W$ is refutable in P, then W is provable in S.

The main consequence of the completeness theorem says that if a first order theory is consistent, then it is satisfiable in a countable domain. Using the method of arithmetic representation of syntax, it is possible to obtain a sharper formulation so that we can, for example, exhibit a relation over natural numbers which serves as an arithmetical translation of the membership relation in set theory[1]:

---

1)   This was worked out in detail in *Methodos*, vol. 3 (1951), pp. 217—232; it makes essential use of the more restricted results in D. Hilbert and P. Bernays, *Grundlagen der Mathematik*, vol. II, 1939.

5.9.4 If we add $Con(S)$ (the usual formula expressing the consistency of a first-order system S) to number theory as a new axiom, we can prove in the resulting system arithmetic translations of all theorems of S.

These results lead to a puzzling phenomenon. If T is an impredicative extension of S, the nit is well known that $Con(S)$ can be proved in T. On the other hand, if $Con(S)$, then S has a model in number theory so that T has a model in the impredicative extension of number theory. Hence, for moderately strong T we can carry out in T this argument to derive $Con(T)$ from $Con(S)$. Hence, we would seem to reach the result that $Con(T)$ is a theorem of T. By Gödel's second theorem, T would be inconsistent. It turns out that even though we can find suitable S and T so that most of the above arguments can be carried out, the whole proof breaks down because certain necessary applications of mathematical induction cannot be proved in T[1].

A result which has been quoted very often in connection with the search for computer proofs of theorems of first order logic is a theorem of Herbrand regarded by him as "a more precise statement of the well-known Löwenheim-Skolem theorem." Using the discussion so far, the theorem can be stated in the following form in which H is a more or less standard formal system for first order logic[2]:

5.9.5 (1a) If a schema W does not have a solution of level k, for some k, then W is refutable in H (i.e., $\neg W$ is provable in H) (1b) Given the number k, we can construct a refutation of W. (2a) If W is refutable in H, then there is a number k such that W has no solution of level k. (2b) There is an elementary method by which, given a refutation of any schema W, we can find a number k such that W has no solution of level k.

We have tried to avoid too much formal detail in our considerations. But perhaps a sample of some of the common explanations might give a flavor of the formally more explicit treatment.

---

1) This was considered at length in *Trans. Am. Math. Soc.*, vol. 73 (1952), pp. 243—275.

2) J. Herbrand, *Recherches sur la théorie de la démonstration*, 1930, thesis at the University of Paris.

For each collection of relations (or predicates), we can set up a corresponding first order language. A simple example is the first order language L_s of set theory which has the single binary membership relation symbolized by $\in$. The language is built up from the following symbols: (i) $\in$; (ii) an infinite set of variables v_0, $v_1\cdots$; (iii) $=$; (iv) $\supset, \neg, \exists$.

If t_1 and t_2 are variables, then $t_1 \in t_2$ and $t_1 = t_2$ are atomic formulas. Formulas are defined inductively: (i) an atomic formula is a formula; (ii) if F and G are formulas, then $F \supset G$ and $\neg F$ are formulas; (iii) if F is a formula and v_i is a variable, then $\exists v_i F$ is a formula.

Sometimes a predicate language may also include symbols for functions and constants standing for distinguished elements. Then atomic formulas are of the form $R_i(t_1, \ldots, t_n)$ where $t_1, \ldots, t_n$ are terms. Terms are defined inductively: (i) a constant is a term; (ii) a variable is a term; (iii) if $t_1, \ldots, t_m$ are terms and f is a symbol for a function with m arguments, then $f(t_1, \ldots, t_m)$ is a term. These additional symbols may be natural and convenient even though it is possible to treat a function with n arguments as a relation of degree $(n+1)$ and a constant a as a unary relation $\{a\}$.

Variables can occur in a formula free or bound. When all variables occurring in a formula are bound in it, the formula is also called a sentence. A variable occurs free in a formula F if it occurs in F and: (i) if F is an atomic formula; or (ii) if F is $G \supset H$, and it occurs free in G or in H; or (iii) if F is $\exists v_i G$ and it occurs free in G and is distinct from v_i. The distinction of free and bound occurrences is interesting especially when we come to interpret a first order language to make its sentences true or false. To interpret a formula containing free variables in (say) the simple language L_s requires a relational structure $\langle A, \varepsilon \rangle$, i.e., a universe A and a relation ε defined over A^2, as well as an assignment of elements of A to the free variables (or a map of the variables into A).

In order to define truth in a relational structure of sentences in a corresponding first order language, it is generally necessary to go through a somewhat confusing detour by saying that certain objects satisfy a formula with free variables. It is easier to grasp this detour if one imagines that every element of the universe A

has a name. Suppose we have chosen $\langle A, \varepsilon \rangle$ and wish to define inductively which sentences of L_s are true. We define first a relation of "satisfying" between finite sequences of objects in A and all formulas in the following manner:

(i) $\langle a_1, a_2 \rangle \models v_i \in v_j$ iff $a_1 \varepsilon a_2$; $\langle a_1, a_2 \rangle \models v_i = v_j$ iff $a_1 = a_2$.

(ii)   If $\neg F$ contains n free variables, then $\langle a_1, \ldots, a_n \rangle \models \neg F$ iff it is not the case that $\langle a_1, \ldots, a_n \rangle \models F$.

(iii)   If $F \supset G$ contains n free variables $v_1, \ldots, v_n$ (say) and F contains a subsequence of them $v_{i_1}, \cdots, v_{i_p}$ and G contains a subsequence $v_{j_1}, \cdots, v_{j_q}$. Then $\langle a_1, \cdots, a_n \rangle \models (F \supset G)$ iff $\langle a_{i_1}, \cdots, a_{i_p} \rangle$ does not $\models F$ or $\langle a_{j_1}, \cdots, a_{j_q} \rangle \models G$.

(iv)   If $\exists v_i F$ contains n free variables, then $\langle a_1, \ldots, a_n \rangle \models \exists v_i F$ iff either v_i does not occur free in F and $\langle a_1, \ldots, a_n \rangle \models F$ or else there is some a_{n+1} in A such that $\langle a_1, \ldots, a_{n+1} \rangle \models F$.

Finally, since a sentence F is a formula with no free variables, we say that F is true in $\langle A, \varepsilon \rangle$ or $\langle A, \varepsilon \rangle \models F$ iff the empty sequence satisfies F according to the above definition[1].

---

1)   For explicit treatments of truth definitions, see Hilbert-Bernays mentioned in footnote on page 90 and the paper under footnote 1) on page 91.

6. COMPUTATION: THEORETICAL AND PRACTICABLE

6.1 Computation in polynomial time

We have considered the rudiments of theoretical computability in terms of Turing machines and recursive functions in Appendix C. In this chapter we shall begin with some discussion of practicable or feasible computation and then move on to some more advanced results on theoretical computability. It is of interest to observe that theoretical computability not only has been studied earlier but also is easier to deal with (thus far at least) than feasible computation.

If we consider counting, addition, multiplication, exponentiation, we see that we take bigger and bigger steps. For example, instead of adding m and m, multiplication adds m any number of times together. This suggests a sequence of functions whose values increase faster and faster. Roughly speaking, the sequence corresponds to what are known as primitive recursive functions[1]. Once we give the exact definition of this sequence of functions, we can define computable functions which increase faster than all of them.

More explicitly, we can keep track of the iterations by an index and introduce a sequence of functions:

$$f_1(n) = n + 1.$$
$$f_2(n, 1) = f_1(n), \qquad f_2(n, m+1) = f_1(f_2(n, m)).$$
$$f_3(n, 1) = f_2(n, n), \qquad f_3(n, m+1) = f_2(f_3(n, m), n).$$
$$f_{k+1}(n, 1) = f_k(n, n), \quad f_{k+1}(n, m+1) = f_k(f_{k+1}(n, m), n).$$

It is easy to convince oneself that for each k, we can compute the values of $f_k(m, n)$ for all m and n. But then we can also define a new computable function $f(m, n) = f_m(n, n)$, which clearly increases faster

---

1) For a detailed consideration of primitive recursive functions and their simple extensions, see R. Peter, *Rekursive Funktionen*, 1957.

than every function in the original sequence. This phenomenon seems to present an obstacle to defining the set of all computable functions. We have seen in Appendix C how this obstacle has been overcome.

When it comes to feasible computation, even the functions in the above sequence, simple as they are, are mostly not computable. One speaks of "exponential explosion", meaning that the exponential function is no longer feasibly computable. Roughly speaking f_4 in the above sequence corresponds to exponentiation so that every f_k, $k \geqslant 4$, in the above sequence is clearly beyond feasible computation. In fact, even when the value M is a fixed but large number, it is clear that the function $f(n) = n^M$ is not feasibly computable. To impress on people how large exponentiation quickly gets, somebody has compiled a table on the time needed for completing so many operations. Suppose a computer can perform 10^6 operations per second and the complexity function $f(n)$ gives the number of operations needed to answer a question of size n. We have:

		size n					
		10	20	30	40	50	60
complexity	n^2	0.0001 second	0.0004 second	0.0009 second	0.0016 second	0.0025 second	0.0036 second
	n^3	0.001 second	0.008 second	0.027 second	0.064 second	0.125 second	0.216 second
	n^5	0.1 second	3.2 seconds	24.3 seconds	1.7 minutes	5.2 minutes	13 minutes
	2^n	0.001 second	1 second	17.9 minutes	12.7 days	35.7 years	366 centuries
	3^n	0.059 second	58 minutes	6.5 years	3,855 centuries	2×10^8 centuries	1.3×10^{13} centuries

The table illustrates the remarkable fact that even though computers can operate very fast, crude methods of computation can quickly get beyond the capacity of computers. It also shows concretely the wide gap between theoretical and feasible computation.

There is no general agreement on a sharp definition of feasible computability. In recent years much attention has been paid to the natural class of polynomial complexity which is mathematically interesting because of its stability. A given infinite set of problems is computable in polynomial time if there is a polynomial f such that for each problem (whose expression is) of length n, the problem is answered in $\leqslant f(n)$ steps (or equivalently, there is a fixed N, such that the problem is answered in $\leqslant n^N$ steps). If there is no polynomial bound, then, for every general method, there are infinitely many n, such that the decision of some problem of length n requires exponential time (e.g., $\geqslant 2^{\sqrt{n}}$). In that case, it is generally accepted that the set of problems has no feasible general solution.

We have illustrated applications of correlations of Turing machines and tiling problems with sentences of first order logic in Appendix C. In 1971, this type of correlation was extended to sentences of propositional logic with a wide range of consequences. In order to give an exposition of this area, we explain first a few terms to be employed.

6.2   The tautology problem and NP completeness

The tautology problem is the problem of deciding quickly whether a sentence of propositional logic (or a Boolean expression) is valid or tautologous. One sharper question is: can the decision be made universally in polynomial time or does every general method require exponential time for an infinite number of special cases? While propositional expressions have been intensively studied, when it comes to the complexity of decision procedures, the area remains wide open and not enough results have been obtained thus far even to warrant a conjecture as to the answer to this general question. For example, we do not know either that it can be done in polynomial time nor that it cannot be done in quadratic time. Hence, the ignorance is pretty complete.

We shall have occasion to consider the tautology problem in two equivalent special forms. First, we can begin with propositional sentences in disjunctive normal form, i.e., a disjunction of conjunctions of literals (i.e., each being propositional letter or its negation), and ask whether they are tautologies; this is called DNF validity problem. Second, we can begin with sentences in conjunctive normal form, i.e., a conjunction of disjunctions of literals, and ask whether they are satisfiable; this is called the CNF satisfiability problem. As we come to special examples, these notions will become clear.

We have considered Turing machines in Appendix C. We shall now make use of nondeterministic Turing machines as well. While an algorithm gives a succession of steps, a nondeterministic machine gives a tree structure and behaves like a proof procedure rather than a decision procedure. A set S of sentences belongs to P if there is a Turing machine which accepts it (in the sense that every sentence, if in S, is recognized to be so) in polynomial time; it belongs to NP if there is a nondeterministic "method" with a polynomial f such that for each sentence in S of length n, there is a path in the solution tree (a shortest proof) which is no longer than $f(n)$.

We note a difference between P and NP. For concreteness, let C be the set of all propositional sentences in CNF and S be the subset of the satisfiable ones. If S is in P, then there is a polynomial f and a Turing machine T such that, for every w in S with length $|w|$, T can yield a positive answer in time $t \leqslant f(|w|)$. It then follows that the complement of S in C is also in P because we can simply use $f(n) + 1$ and modify the machine T slightly to say that w belongs to $C - S$ iff w has not been recognized to be in S at time $f(|w|)$. Hence, for sets in P we can equivalently think of a polynomial decision procedure in each case, to decide, for example, whether an input w in C belongs to S or not. If we are trying to show that S is in P, then it is equivalent to the tautology problem.

But the situation with NP is quite different. It is not known generally whether the complement of a set in NP is also in NP. For example, it is easy to show (see below after 6.2.1) that the set of satisfiable sentences is in NP, but it is not known that

(I)   the set of tautologies (valid sentences) is in NP.

In fact it has been shown that if (I) is true, then NP is closed with respect to complementation; and also that if the set of tautologies is not in NP, then the set of satisfiable sentences is not in P, so that $P \neq NP$[1]. Hence, we see further ramifications of the tautology

1)   S. A. Cook and R. A. Reckhow, *Proc.* 6th Annual ACM Conference on Theory of Computing, 1974, pp. 15—22.

problem, which, however, can be disregarded by those who work toward showing that the set of tautologies is in P.

A set C of sentences is said to be P-reducible (polynomial reducible) to a set D of sentences if assuming a code of members of C and D into numbers, there is a polynomial transformation f of sentences in C into those in D so that for all q in C, q gets a positive answer iff $f(q)$ does. We shall leave out more refined distinctions of reducibility. A set C of sentences is said to be NP complete iff C belongs to NP and all other sets in NP are P-reducible to C. Let Satis be the set of satisfiable sentences of the propositional logic (or, equivalently those in CNF).

6.2.1   The set Satis is NP complete.

This result has led to the discovery of various NP complete sets of problems. Since the class NP includes many problems of practical importance, one gets the conclusion that a polynomial algorithm for any NP complete problem would yield polynomial algorithms for all other NP problems[1].

Certainly the set of satisfiable sentences belongs to NP. Let H be given, containing k propositional letters $p_1, \ldots, p_k$. Let H_0 (and H_1) be obtained from H by putting true (and false) for p_1, H_{00} (and H_{01}) be obtained from H_0 by putting true (and false) for p_2, etc. In this way, we arrive at a full binary tree with $k+1$ levels. But H is satisfiable iff there is one final node in this tree which gets the value true. With a nondeterministic machine, we can always assume that such a branch is chosen (if there is one).

Assume now K is a set of sentences in NP, we want to show that K is P-reducible to Satis. In other words, there is a nondeterministic machine T and a polynomial $Q(n)$ such that for every $w \in K$, w is accepted by T in time $Q(|w|)$. We wish to define a function S so that $S(w)$ is a sentence in propositional logic and that $S(w)$ is satisfiable iff $w \in K$.

---

1)   Theorem 6.2.1 was proved by S. A. Cook in ''The complexity of theorem proving procedures,'' *Proc.* 3rd Annual ACM Conference on Theory of Computing, May, 1971, pp. 151—158. Many more NP and NP complete problems are given by R. M. Karp in *Complexity of computer computations*, ed. R. E. Miller and J. Thatcher, 1972, pp. 85—104.

Let $s_1, \ldots, s_m$ be the alphabet of T and $q_1, \ldots, q_n$ be the states of T. We use a one-way infinite tape and number the squares 1, 2, 3, etc. Note that for each w, no square beyond $Q(|w|) = t_w$ ever gets scanned.

We introduce a lot of propositional letters for each given w.

(1) $p(i, s, t)$, $1 \leqslant i \leqslant m$, $1 \leqslant s$, $t \leqslant t_w$. Intuitively, $p(i, s, t)$ is true iff square s at time t contains s_i.

(2) $q(i, t)$, $1 \leqslant i \leqslant n$, $1 \leqslant t \leqslant t_w$. Intuitively, $q(i, t)$ is true iff at step t, T is in state q_i.

(3) $r(s, t)$, $1 \leqslant s$, $t \leqslant t_w$. Intuitively, $r(s, t)$ is true iff at time t, square s is being scanned.

With the help of these letters, we can express the operations of T fairly directly but lengthily.

(4) B asserts that at each step t, exactly one square is scanned.

$$B(t) = (r(1, t) \vee \cdots \vee r(t_w, t)) \wedge$$
$$[\wedge (1 \leqslant i < j \leqslant t_w)(\neg r(i, t) \vee \neg r(j, t))]$$
$$B = B(1) \wedge \cdots \wedge B(t_w).$$

(5) C asserts that each square s at each time t contains exactly one symbol.

(6) D asserts that at each time t, T is at exactly one state.

(7) E asserts the initial condition for T, containing initial state, initial tape with w on it and remainder (up to square t_w) all blank, as well as the square initially under scan.

(8) F, G, and H assert that the letters under (1), (2), and (3) are properly updated in accordance with the state tables of T. Observe that at each state, there are in general several possible next states because T is non-deterministic. Suppose T is at t in state q_i scanning s_j and $q_{i_1}, \cdots, q_{i_k}$ are the possible next states.

$$G(i, j) = \wedge (1 \leqslant s, t \leqslant t_w - 1)(\neg q(i, t)$$
$$\vee \neg p(j, s, t) \vee \neg r(s, t) \vee q(i_1, t + 1) \vee \cdots \vee q(i_k, t + 1))$$
$$G = \wedge (1 \leqslant i \leqslant n, 1 \leqslant j \leqslant m)G(i, j).$$

(9) A asserts that T accepts w: simply $q(a, t_w)$, where q_a is the unique accepting state and we assume that once T enters q_a, it remains in it forever.

Let now $S(w)$ be $A \wedge B \wedge C \wedge D \wedge E \wedge F \wedge G \wedge H$.

We wish to argue that $S(w)$ is satisfiable iff $w \in K$. Suppose $w \in K$. There is then a short computation which begins with E and ends with A. This yields a truth assignment which satisfies $S(w)$ according to the intuitive explanation of the propositional letters under (1), (2), and (3). Suppose $S(w)$ is satisfiable. There is then a satisfying truth assignment from which we can copy out a computation that accepts w.

6.2.2   *Corollary*.   If Satis $\in$ P, then P = NP.

Since any set K in NP is P-reducible to Satis, by the definition of P-reducibility, K is in P if Satis is. We note incidentally that in the proof of 6.2.1 (and 6.2.2), we can also take Satis in CNF.

This corollary and the fact that there are many interesting problems which are in NP have stimulated much interest in the satisfiability problem and the tautology problem. From the considerations so far, we can summarize the general open problems and their relation to these special problems on propositional sentences (Boolean expressions).

6.2.3   Two general open problems are: is P = NP? is NP closed under complementation? If P = NP, then NP is closed under complementation, but the converse is not known.

6.2.4   The special problems are: is Satis in P? is the set of tautologies Taut in P? Is Taut in NP? It is known that Satis $\in$ NP. If Satis $\in$ P, then P = NP. If Taut $\in$ NP, then NP is closed under complementation. If Taut is not even in NP, then Satis is not in P and P $\neq$ NP.

In a later section, we shall report on some attempts to show that both Satis and Taut are in P. For such a purpose, there is no need to treat Satis and Taut symmetrically.

6.3   Examples of NP problems

There is a long list of NP complete problems[1]. We mention just a few. The set HP of Hamiltonian paths discussed in Chapter 4, section 4.5, is NP complete. Another related NP complete set is the

---

1)   See footnote on page 98.

set NC of G such that G is a finite graph which contains a set R of m nodes covering G in the sense that every edge in G is incident to some node in R. A third NP complete set is the set CL of G such that the finite graph G has k adjacent nodes. It is fairly simple to show that these sets are all in NP. It can also be shown that CL is P-reducible to NC and NC is P-reducible to HP. We merely outline for illustration a proof that Satis is reducible to CL.

The idea is to find for each sentence A in CNF a corresponding G_k such that A is satisfiable iff G_k has a clique of size k. Let A be $(L(1, 1) \vee \cdots \vee L(1, n_1)) \wedge \cdots \wedge (L(m, 1) \vee \cdots \vee L(m, n_m))$, where each $L(i, j)$ is a literal. Define a graph as follows. For each pair $\langle L(i, j), i \rangle$, $1 \leqslant i \leqslant m$, $1 \leqslant j \leqslant n_i$, define a node. Connect the nodes by the following criterion: $\langle L(i, j), i \rangle$ be adjacent to $\langle L(u, v), u \rangle$ iff $i \neq u$ and $L(i, j)$ is not the opposite of $L(u, v)$ (the opposite of p is $\neg p$ and conversely). Let $k = m$.

If A is satisfiable, there is a truth assignment such that at least one literal in each disjunction clause gets the value true. The set of k nodes corresponding to these m literals (recall $k = m$) forms a clique because any two of them are in different clauses and cannot be the opposite of each other. Conversely, a clique of $k(= m)$ adjacent nodes determines a satisfying assignment for A.

The following sets are examples which are in NP but not known to be NP complete: the set of prime numbers, the set of composite numbers, the set of isomorphic graph pairs. In fact, the proof that the set of primes is in NP is rather involved.

6.4 The tautology problem

We digress to report on some detailed work aimed at deciding propositional expressions quickly[1]. In this section only, we shall use $\bar{A}$ in place of $\neg A$ for the purpose of brevity.

The general problem is to decide whether an arbitrary truth-functional expression is valid (or satisfiable or contradictory). The familiar connectives are: not, and, or, only if, exclusive or, if and

---

1) This section is extracted from the paper with B. Dunham, *Annals of math. logic*, vol. 10 (1976), pp. 117—154. References are given in that paper to related earlier works.

only if ($\equiv$). Of these, "p only if q" is the same as "not p or q"; "p (exclusive) or q" is the same as "$p \equiv \bar{q}$". Hence, we may confine our attention to not, and, or, $\equiv$. It is common to eliminate $\equiv$ also, but this is not always desirable because $\equiv$ has many elegant properties and its elimination gives way to quite complex expressions in terms of not, and, or. Given the choice of connectives, there is also the question of the extent to which normal forms are to be used in our investigation.

The two most familiar normal forms are the conjunctive and the disjunctive, in terms of not, and, or. A first observation is that the validity of an expression in conjunctive normal form can always be tested quickly because it is valid only if every disjunction in it is valid (i.e., there is some variable p such that both p and $\bar{p}$ appear in it). Symmetrically, satisfiability of an expression in disjunctive normal form can be tested quickly. We recall that an expression A is valid if and only if its negation $\bar{A}$ is not satisfiable (i.e., contradictory). Rather the open problem for expressions in conjunctive (resp. disjunctive) form is to test quickly whether it is satisfiable or contradictory (resp. valid). This familiar observation illustrates the fact that we cannot assume without additional argument that expressions are given in any normal form, because, for example, the question of testing validity quickly of arbitrary expressions is reduced to the question whether converting an arbitrary expression into the conjunctive normal form can be done in polynomial time.

It has been pointed out that we can, by suitably introducing new variables, quickly turn an expression A into an expression B in disjunctive form such that A is valid if and only if B is. Therefore, if we are studying validity, we can confine our attention to expressions in disjunctive form. Similarly, we can assume expressions turned into conjunctive normal form when studying satisfiability.

The widely studied resolution method is based on the familiar "cut rule": from $B \lor \bar{p}$ and $p \lor C$, infer $B \lor C$. When we begin with an expression A in conjunctive normal form, we can test whether A is contradictory by making all possible applications of the cut rule to all clauses at each stage until either we reach two conclusions p and $\bar{p}$ for some variable p or, failing that, can no longer apply the cut rule to get any new clauses. This method appears to be quite fast and generally it is not easy to find examples which cannot be decided quickly by familiar methods. Such counterexamples are useful in stimulating the search for more efficient methods.

In 1968 a set of "grid problems" was proposed and shown to be requiring exponential time for decision by the resolution method. Afterwards a set of "occupancy problems" was found to have the same property. We shall omit the description of the grid problems but merely give a method by which they can be decided quickly[1].

For each n, the occupancy problem of level n says intuitively that n objects cannot occupy all of $(n+1)$ drawers. We express this fact by a Boolean expression in disjunctive form as follows.

6.4.1 *The occupancy problems.*

Let p_{ij} say intuitively that drawer i contains object j ($1 \leqslant i \leqslant n+1$, $1 \leqslant j \leqslant n$). For each n, the expression O_n is a disjunction of the following clauses: For $1 \leqslant i \leqslant j \leqslant n+1$, $1 \leqslant k \leqslant n$, $p_{ik}p_{jk}$ (drawer i and drawer j both contain object k). For $1 \leqslant i \leqslant n+1$, $\bar{p}_{i1} \cdots \bar{p}_{in}$ (no object is in drawer i). (Intuitively, $\exists i \exists j \exists k p_{ik}p_{jk} \vee \exists i \forall n \bar{p}_{in}$.)

In other words, either there are two drawers sharing one object or there is an unoccupied drawer. It is easy to see that O_n contains $n(n+1)$ variables and $(n+1)(n^2+2)/2$ clauses. It has been calculated that by the routine application of the resolution method the time for deciding O_n is exponential: $n(n+3)2^{n-2}$.

6.4.2 The special case O_3 is, for example, a disjunction of the following twenty-two clauses:

(1) $\bar{a}\bar{b}\bar{c}$, $\bar{d}\bar{e}\bar{f}$, $\bar{u}\bar{v}\bar{w}$, $\bar{x}\bar{y}\bar{z}$;   (2) ad, au, ax, du, dx, ux;

(3) be, bv, by, ev, ey, vy;   (4) cf, cw, cz, fw, fz, wz

By combining a method of inversion with a concept of "giant terms", it is possible to decide quickly the occupancy problems from general considerations.

We observe that if an expression in disjunctive form does not contain any absolutely positive clauses (or similarly negative ones),

---

1) The grid problems were introduced by G. S. Tseitin in *Studies in constructive math. and math. logic*, Part II, ed. A. O. Slisenko, 1968, pp. 115—125. They are restated in the paper listed under the preceding footnote. The occupancy problems were introduced by R. M. Karp and S. A. Cook.

then it cannot be valid. For example, if there is no clause in which no variable is negated, then by assigning every variable the value true, the whole expression must get the value false. In any expression A if we "invert" any variable by substituting a variable p for $\bar{p}$ and $\bar{p}$ for p, then the result is valid if and only if A is. Hence, we can sometimes try to invert suitable variables to eliminate absolutely positive clauses (or absolutely negative ones). Of course, if the expression is valid, we cannot find such inversions. And it is possible to establish that such inversions do not exist. This yields a decision method which is not efficient in the general case but is efficient for certain interesting special cases.

The inversion method is closely related to the concept of giant terms.

6.4.3   Let A be an expression in disjunctive form. A disjunction G of literals is called a giant term for A if and only if G contains at least one literal from each clause of A and does not contain both p and $\bar{p}$ for any p.

6.4.4   An expression A in disjunctive form is valid if and only if there is no giant term for it.

We omit the fairly direct proof of 6.4.4.

6.4.5   Let A be an expression in the disjunctive form. There is a giant term for A (i.e., A is not valid) if and only if there is an inversion of A which contains no absolutely positive clause.

Suppose there is a giant term G for A, say $\bar{p} \vee q \vee \bar{s} \vee t$. That means A contains no clause which makes $p\bar{q}s\bar{t}$ true. Suppose a, b, c, d are all the other variables in A. Then A does not cover the row in the truth table which makes $abcdp\bar{q}s\bar{t}$ true. Therefore, if we invert q and t, the result obtained can contain no absolutely positive clause, because any such clause would cover the missing row.

Conversely, suppose there is an inversion of A, say q to $\bar{q}$ and t to $\bar{t}$, which yields a disjunction B that contains no absolutely positive clauses. Suppose a, b, c, d, p, q, s, t are the only variables in A. Then $\bar{a} \vee \bar{b} \vee \bar{c} \vee \bar{d} \vee \bar{p} \vee \bar{q} \vee \bar{s} \vee \bar{t}$ must be a giant term for B, since every clause in B contains at least one negative literal. But then $\bar{a} \vee \bar{b} \vee \bar{c} \vee \bar{d} \vee \bar{p} \vee q \vee \bar{s} \vee t$ would be a giant term for A.

By taking advantage of certain special features of the occupancy problem, we can decide all its cases efficiently with the method of inversion and the related concept of giant terms. Consider, for illustration, the simple case O_3 mentioned above. A first mechanically recognizable feature is that the clauses in the expression can be separated into groups containing different literals:

O_3. A_1: $ad,\ au,\ ax,\ du,\ dx,\ ux$;

   A_2: $be,\ bv,\ by,\ ev,\ ey,\ vy$;

   A_3: $cf,\ cw,\ cz,\ fw,\ fz,\ wz$;

   B_1: $\bar{a}\bar{b}\bar{c}$; B_2: $\bar{d}\bar{e}\bar{f}$; B_3: $\bar{u}\bar{v}\bar{w}$; B_4: $\bar{x}\bar{y}\bar{z}$.

No two clauses in any two different groups have any literal in common.

We now ask whether there is any giant term for O_3. There are altogether 12 variables. Therefore, a giant term must contain no more than 12 literals. Consider A_1 first. It is easy to see that at least three variables must occur in a giant term. Similarly with A_2 and A_3. Hence, we need 9 unnegated variables for A_1, A_2, A_3. But each of B_1, B_2, B_3, B_4 needs one additional literal since all literals in them are negative. Therefore, we need at least 13 literals to make a giant term, which is impossible. Hence, O_3 is valid.

Observe that if we delete any clause, say ux, we would have a giant term, viz. the disjunction of $a,\ d,\ b,\ v,\ y,\ f,\ w,\ z,\ \bar{c},\ \bar{e},\ \bar{u},\ \bar{x}$. Therefore, the result obtained from O_3 by deleting any single clause is not valid.

It is not hard to convince oneself that the decision procedure can be mechanized and it applies efficiently to all cases of the occupancy problem.

There is a quick decision procedure for a class of expressions involving primarily the connective $\equiv$. This method decides quickly the class of grid problems mentioned before. We give a description of this method.

6.4.6   The biconditional is commutative and associative; hence, we can use the notation $A_1 \equiv \cdots \equiv A_n$ for a chain of biconditionals. Of course, this is analogous to disjunction and conjunction rather than $=$; for example, $A \equiv B \equiv C$ does not mean that $A \equiv B$ and $B \equiv C$.

6.4.7   In any chain of biconditionals, we can move any negation sign governed directly by $\equiv$ to govern the whole chain; hence, any even number of negation signs can be dropped (for example $\bar{p} \equiv \bar{q} \equiv \bar{n}$ is equivalent to $p \equiv q \equiv \bar{n}$ or $p \equiv \bar{q} \equiv n$ or $\bar{p} \equiv q \equiv n$) so that at most one term of the chain begins with a negation sign.

6.4.8   Whenever a literal occurs twice in a chain of biconditionals, both occurrences can be dropped; hence, no term has to occur more than once (in a chain of length ≥ 3).

6.4.9   An expression is in (single literal conjunctive) biconditional normal form if it is a conjunction of clauses each of which is single literal or a chain of biconditionals in which each term is a literal.

We do not know whether each expression can be turned quickly into such a normal form, but once an expression is in this form, a general method can be applied to decide quickly whether it is satisfiable. Take any expression A in this biconditional form. Suppose it contains n variables $p_1, \ldots, p_n$. By 6.4.7 and 6.4.8 we can simplify A so that each clause contains no more than n terms with no more than one variable negated. Let B be the simplified expression. If A or B contains any clause $p_i \equiv p_i$, it can be dropped; if A or B is simply $p_i \equiv p_i$, then B and A are of course satisfiable. If A or B contains a clause $p_i \equiv \bar{p}_i$, then B (and therewith A) is not satisfiable and we are through. Therefore, we can assume there are no such clauses and each clause contains each variable at most once. A or B may also contain clauses of the form p_i or $\bar{p}_i$. If for some i, A or B contains p_i and $\bar{p}_i$, we are through because B (and therewith A) is not satisfiable.

In general, B contains chains of biconditionals and possibly single literal clauses. Consider all chains containing p_1, if there are any. They are (say) $p_1 \equiv A_1, \ldots, p_1 \equiv A_m$.

6.4.10   The conjunction of $p \equiv A_1, \ldots, p \equiv A_m$ is satisfiable if and only if the conjunction of $A_1 \equiv A_2, \ldots, A_{m-1} \equiv A_m$ is.

Suppose an assignment is given which makes $p \equiv A_1, \ldots, p \equiv A_m$ all true. p is assigned 0 or 1. In either case, $A_1, \ldots, A_m$ all get the same value. On the other hand, suppose $A_1 \equiv A_2, \ldots, A_{m-1} \equiv A_m$ has a true assignment. Then $A_1, \ldots, A_m$ all get the same value. If we give p that value, then $p, A_1, \ldots, A_m$ all get the same value, and therefore $p \equiv A_1, \ldots, p \equiv A_m$ are all true.

Hence, we can replace the m clauses $p_1 \equiv A_1, \ldots, p_1 \equiv A_m$ by the

$(m-1)$ clauses $A_1 \equiv A_2, \ldots, A_{m-1} \equiv A_m$, which can again be simplified by applying 6.4.7 and 6.4.8. Observe that the resulting clauses are each at most of length $n-1$, and therefore the increase in length is well under control. In the process, we may generate single literal clauses or $p_i \equiv p_i$ or $p_i \equiv \bar{p}_i$. These are treated as before. Now that we have eliminated p_1, except possibly in a single literal clause, with little expansion, we can continue with the other variables in the same manner. In the final result (if we have not reached a decision earlier), we can have only clauses of the following four forms: p_i, $\bar{p}_i$, $p_n \equiv p_n$, $p_n \equiv \bar{p}_n$. If there is a clause of the form $p_n \equiv \bar{p}_n$, then the original expression A is not satisfiable; similarly if there are two clauses p_i and $\bar{p}_i$ with the same i. Otherwise, we can assign 1 to p_j for each clause of the form p_j and 0 to p_k for each clause of the form $\bar{p}_k$. It is then seen that A is satisfiable.

6.5 Polynomial time and feasibility

As we have noted before, it is not even known that the tautology problem cannot be decided in quadratic time. The interest in polynomial time is understandable because we have here a natural domain which is closed under many operations and, therefore, rather easy to work with. It is, however, doubtful that a polynomial $Q(n)$ of very high degree could be accepted as the bound for feasible computations. Also, it is generally true that when we do find polynomial bounds on the time needed, we need only polynomials of rather low degrees.

If we agree that polynomial time is too broad a class to capture the idea of feasibility, then we are hard put to find other natural stopping places. In fact, some people feel that we can never find any uniform simple mathematical characterization of what feasible computations are.

One attraction of polynomial time is the fact that so many different sets of computation problems are seen to be of the same order of difficulty. A natural question to ask is whether some narrower classes would also share to some extent this property. Another suggestion is to have different degrees of feasibility: for example, to say that certain programs are, even though feasible, not efficient.

At any rate, it seems clear that at present there is much we do not understand about fast decision procedures, fast proof methods, efficient computer programs, etc. There are many different directions along which people are pursuing. We content ourselves with mentioning as illustration one recent proposal in terms of quasilinear time bounds[1].

Consider computations on multitape (deterministic or nondeterministic) Turing machines. Time bounds of the orders $n(\log n)^k$ for fixed values of k are called quasilinear in n. Let QL and NQL be the quasilinear analogues of P and NP for polynomial time. For example, integer multiplication, sorting, and searching can be done in quasilinear time. Moreover, Satis in not only in NQL but also NQL complete. Analogues of results on P and NP have been proved. For example:

6.5.1   QL $=$ NQL iff Satis $\in$ QL.

If QL $=$ NQL, then Satis $\in$ QL and certainly Satis $\in$ P. Hence, if QL $=$ NQL, then P $=$ NP but the converse is not clear. It would, therefore, seem plausible to suggest that if one believes P $=$ NP, it would be easier to work with P and NP, but that if one believes P $\neq$ NP, it would be easier to work with QL and NQL. In fact, it is, one certainly feels, an easier task to show that Satis is in P rather than in the much narrower class QL. Symmetrically, to show that Satis is not in QL seems easier than to show that it is not in the much broader class P.

6.6   Decidable theories and unsolvable problems

We have mentioned in 3.5 and 3.8 how the study of sequential circuits had led to the formulation and decidability of new theories, as well as some positive and negative results on how fast some of the familiar decidable theories can be decided.

There are now many results on different decidable classes and theories which cannot be decided in polynomial time including the familiar areas of monadic first order logic and ordinary arithmetic

---

1) C. P. Schnorr, *J. ACM*, vol. 25 (1978), pp. 136—145. References to related works are given in the paper. Compare also A. V. Ako, J. E. Hopcroft, and J. D. Ullman, *The design and analysis of computer algorithms*, 1974.

with addition only ("the Presburger arithmetic"). The general idea of such proofs is related to the method of proving the incompleteness theorems and the device of correlating sentences with behaviors of machines. Computations on Turing machines are coded and short sentences are constructed to assert that the coded computation will eventually halt, but only after a long time. Hence, we find short sentences which can only be decided after a long time. Since each algorithm can be realized on a Turing machine and since the construction can be made in a uniform manner for all such machines, the results apply to all decision methods. This is analogous to extending "I am not provable in S" to "I am not provable in S with less than $f(n)$ steps," where n could be the length of the sentence itself and f could be an exponential or worse function. The earliest result in this class seems to have been found in May, 1972[1].

We shall say no more about practicable computations but report briefly on some aspects of the more traditional subject of theoretical computability in the remainder of this chapter.

Up to 1940, the best known results on the decidability of theories are that arithmetic with addition and multiplication is undecidable, while arithmetic with addition only or with multiplication only is decidable. Since that time there are studies to determine how weak a theory remains undecidable[2]. More substantially, many decidable and undecidable theories are discovered. In a more general direction, many (recursively) unsolvable problems have been found.

The methods employed in showing a theory decidable have been loosely classified as including the method of eliminating quantifiers, the model theoretic methods, and the method of relative interpretation or translation. Some of the best known decidable theories are the "elementary" theory of real numbers, the theory of p-adic fields,

---

1) A. R. Meyer, *Notices Am. Math. Soc.*, vol. 19 (1972), A-598. For more discussions on this class of results, see M. J. Fisher and M. O. Rabin's paper in *Complexity of computation*, ed. R. Karp, *SIAM-AMS Proc.*, vol. 7, Am. Math. Soc., 1974, pp. 27—41; A. R. Meyer, *Proc. Int. Cong. Math.* 1974, pp. 477—482; and Chapter C3 in *Handbook of math. logic*. The major part of the last item is devoted to theoretical decidability.

2) An earlier standard work on this question is A. Tarski, A. Mostowski, and R. M. Robinson, *Undecidable theories*, 1953.

the theory of the class of all finite fields, the theory of linearly ordered sets, the theory of commutative groups, and the theory of the class of all Boolean algebras.

We have mentioned some unsolvable problems in Chapter 4 and Appendix C. We shall quote without even giving the definitions of the concepts involved a number of interesting results in this area and refer the reader to the literature[1] for more information.

6.6.1   A group is recursively presented iff it is isomorphic to a subgroup of a finitely presented group.

This theorem of G. Higman (1961) shows the equivalence of the recursion-theoretic notion of recursive presentation with a purely algebraic notion. The unsolvability of the word problem of groups mentioned in Chapter 4 is an easy consequence of this.

6.6.2   The homeomorphy problem for manifolds of dimension ≥ 4 is unsolvable (1958).

The solution of the two-dimensional case goes back to Riemann. For three-dimensional manifolds the problem remains open.

6.6.3   There is no algorithm by means of which we can test a given context-free grammar to determine whether or not it is ambiguous.

This has relevance to computer languages which often use context-free grammars[2].

6.6.4   Every recursively enumerable relation is Diophantine.

This is the general theorem from which the unsolvability of Hilbert's tenth problem is derived.

6.7   Tiling problems

We have given some elementary details on the tiling (domino) problems in Appendices A and C. We shall list here some of the applications of these problems to illustrate how a simple device, by being stable and transparent, can lead to fairly extensive developments. The initial occasion for introducing these tiling problems was

---

1)   See Chapter C2 of *Handbook of math. logic,* both for an explanation of the list of results and for references containing more complete proofs of them.

2)   See J. Hopcroft and J. Ullman, *Formal languages and their relation to automata,* 1969.

to study the ∀∃∀ case of the first order logic sentences. As mentioned in Chapter 5, these tools did settle the question and even gave a simple unification of the whole reduction problem of first order logic relative to the prenex form classification[1]. In fact, these studies have led to results on more refined classifications of reduction problems and decision problems of first order logic[2].

In section 4 of Appendix A we have listed (in footnote on page 101) some intuitively appealing papers having to do with the tiling problems and reported (in footnote on page 103) on the interest in finding small solvable tile sets without periodic solutions by people from diverse backgrounds. The current record is either 16 or even 12. Another problem proposed initially is whether the problem of deciding the solvability of arbitrary tile sets is solvable. This includes the origin-constrained problem and the unrestricted problem. The former was shown to be unsolvable in 1960 (see a proof in Appendix C). The latter was shown to be unsolvable a few years later by Berger[3].

In 1974, Hanf showed that there is a solvable tile set which has no recursive solutions, under the origin-constraint and Meyers extended the result to the unrestricted case[4]. Recently Robinson showed the unsolvability of the origin-constrained tiling problem on the hyperbolic plane, but "I still do not know how to prove undecidability for the unconstrained problem, or how to find tiles which force periodicity".[5] Presumably one can also ask for solvable sets with no recursive solutions on the hyperbolic plane.

---

1) For further discussions of these matters, see, for example, H. Hermes, *Selecta mathematica* II, 1970, pp. 114—140; *Logic colloquium 1969*, 1971, pp. 307—309.

2) For example, the paper with B. Dreben and A. S. Kahr, *Bulletin AMS*, vol. 68 (1962), pp. 528—532; S. O. Aanderaa and H. R. Lewis, *J. symbolic logic*, vol. 39 (1974), pp. 519—548; S. Ju. Maslov, *Trudy Mat. Inst. Steklov.*, vol. 98 (1968), pp. 26—87 (see §12. 3).

3) See Robert Berger, *Memoir AMS*, no. 66, 1966, 72 pp. Further developments include R. M. Robinson, *Invent. math.*, vol. 12 (1971), pp. 177—209 and Yu. Sh. Gurvits and Korjakov, *Sib. math. journal*, vol. 13(2), 1972, pp. 459—463.

4) William Hanf and Dale Meyers, *J. symbolic logic*, vol. 39 (1974), pp. 283—285, 286—294.

5) R. M. Robinson, *Inventiones math.*, vol. 44 (1978), pp. 259—264.

Hanf has used such tiling problems to settle some rather central open problems in mathematical logic, proving that every (recursively) axiomatizable theory is isomorphic to a finitely axiomatizable theory and that there exists a finitely axiomatizable theory with countably many complete extensions. Meyers has shown that there is a solvable set which has no solution that is a Boolean combination of r.e. sets (i.e., bounded trialand-error computable) although it is known generally that every solvable set must have a "trial-and-error computable" solution [1]. There has also been a recent attempt to introduce a measure of complexity to the wide class of problems which can be viewed as tiling problems [2].

6.8 Recursion theory: degrees and hierarchies

In terms of "pure" mathematical logic, the study of computability is a part of recursion theory which tends to make generalizations of the basic concepts of recursive and r. e. (recursively enumerable) introduced in Appendix C. We have mentioned reductions on several occasions (such as P-reducible, etc.). A natural idea is to give some ordering (partial or total) of the complexity of sets (of numbers or of problems) by using concepts of reducibility. Thus if A is reducible to B, then B is at least as complex as A. If in addition B is not reducible to A, then B is more complex than A. This simple idea becomes technical as we come to give precise definitions of reducibility.

The main trend in the systematic study of recursion theory has been for many years to take recursiveness as a unit rather than to break up recursive sets and functions into hierarchies of simpler objects. It is, therefore, natural to think in terms of recursive reducibility and introduce in terms of it different degrees of unsolvability [3].

---

1)   Hanf, *Bulletin AMS*, vol. 81 (1975), pp. 587—589; Meyers and Hanf, *Notices AMS*, vol. 24 (1977), p. A-451.

2)   H. R. Lewis presented a paper "A measure of complexity for combinatorial decision problems of the tiling variety" to the Waterloo Conference on Theoretical Computer Science, August 1977.

3)   A fairly comprehensive early treatment of recursion theory is Hartley Rogers, *Theory of recursive functions and effective computability*, 1967. An even

As we try to make the concept precise, we get several different versions. The basic definitions were introduced by Turing in 1939 and by Post in 1944[1].

Turing speaks of an oracle which would answer the question whether a given sentence A (say Goldbach's conjecture) is true so that an oracle machine would behave differently at some stage according as A is true or false. This leads to the concept of relative recursiveness. A total function f is said to be recursive in a total function g, iff there is a partial recursive function h such that $f(x) = h(g, x)$. In that case f is said to be Turing reducible to g, or $f \leqslant_T g$. The equivalence classes under $\equiv_T$ ($f \equiv_T g$ is defined as $f \leqslant_T g$ and $g \leqslant_T f$) are called degrees of unsolvability. Thus the degrees of unsolvability are partially ordered according to just how unsolvable they are. Of course all recursive functions (and sets) are of the same degree. Since there are r. e. sets which are not recursive, the degree of any such set is different from that of the recursive sets.

Post introduced one-one and many-one reducibility.

6.8.1 For two sets (of natural numbers) A and B, $A \leqslant_m B$ or A is many-one reducible to B if there is some general recursive function f such that $x \epsilon A \equiv f(x) \epsilon B$. If f is one-one, then A is one-one reducible to B, or $A \leqslant_1 B$.

Here we get three different notions of reducibility and it is clear that if $A \leqslant_1 B$ then $A \leqslant_m B$, and if $A \leqslant_m B$, then $A \leqslant_T B$.

Post's paper contains a number of elegant new concepts and problems. We give some examples.

6.8.2 Simple sets exist: a simple set is an r. e. set whose complement is infinite but contains no infinite r. e. subset.

6.8.3 Creative sets exist: a creative set is an r. e. set A for which there exists a partial recursive function f such that whenever the r. e. set W_x coded by x is included in the complement $\bar{A}$ of A, $f(x)$ belongs

---

earlier less comprehensive introduction is Martin Davis, *Computability and unsolvability*, 1958. Classical papers in the area have been collected in Martin Davis, *The undecidable*, 1965 (which will be referred to briefly as Davis 1965). On degrees there is J. R. Shoenfield's monograph, *Degrees of unsolvability*, 1971. Part C of *Handbook of mathematical logic* is devoted to recursion theory and contains 8 chapters with a lot of ramifications.

1) See the papers of these dates by Turing and E. L. Post in Davis 1965.

to $\bar{A} - W_x$.

6.8.4   The following are equivalent: (i) A is creative; (ii) A is an r. e. set to which all r. e. sets are $\leqslant_m$ reducible; (iii) replace $\leqslant_m$ by $\leqslant_1$ in (ii).

The most famous problem raised by Post in 1944 is whether there can be two nonrecursive r. e. sets of different degrees of unsolvability. For several years there had been a good deal of interest in attacking this problem until it was answered in the positive[1] in 1956.

As we have just mentioned, the degree of a set A of natural numbers is simply the equivalence class deg (A) under the relation $\equiv_T$. A set A is r. e. in a set B if A is the range of a function f which is recursive in B. A jump operation is defined on degrees so that for each degree d, its jump d' is the largest degree of sets which are r. e. in sets belonging to d. Usually with the help of the jump operation, the partially ordered set D of all degrees has been studied extensively. The study of degrees seems to be appealing only to some special kind of temperament since the results seem to go into many different directions. Methods of proof are emphasized to the extent that the main interest in this area is said to be not so much the conclusions proved as the elaborate methods of proofs.

Let us now move from computability to definability. The first order theory PA of number theory has ω with the familiar interpretation of addition and multiplication as the intended model N. A set A is definable in N by a formula $F(n)$ iff, for every n,

$$n \in A \text{ iff } N \models F(n).$$

6.8.5   The arithmetical sets are just those sets of natural numbers which are definable in N by first order formulas; a set of natural numbers is called an analytical set if it is definable in the structure N by a second order formula.

The arithmetical sets constitute a subclass (of more easily definable sets) of the class of analytical sets. A natural question is to find some suitable measurement of how definable a given set is. The usual

---

1)   Independently by Richard Friedberg, *Proc. Nat. Acad. U.S.A.*, vol. 43 (1957), pp. 236—238 and A. A. Mucnik, *Doklady Akad. Nauk S.S.S.R.*, vol. 108 (1956), pp. 194—197.

starting point is to begin with recursive sets regarding them as of the same complexity in terms of definability. Even though we begin with only addition and multiplication in N, it is known from Gödel's 1931 paper that all recursive sets (relations) are arithmetical because they can all be defined from addition and multiplication with the help of quantifiers (over natural numbers).

The next natural idea is to consider the move from recursive relations to r. e. relations. An r. e. set A is just $\{x \mid \exists y Rxy\}$ for some recursive relation R. Hence, all r. e. relations are also arithmetical. More generally the complement of any arithmetical relation is arithmetical and the projection of one (i.e., a relation r. e. in an arithmetical relation) is arithmetical since first order formulas are closed with respect to $\neg$ and $\exists$.

Since every first order formula is equivalent to one in the prenex form, we limit our attention to such formulas. The choice of taking recursive relations as the simplest has the consequence that, by familiar simple devices, a string of quantifiers of the same kind (i.e., all $\exists$'s or all $\forall$'s) can be contracted equivalently into one. This leads to the following definition. Let $\Pi_0 = \Sigma_0$ be the class of recursive relations, Σ_{n+1} be the class of all projections of Π_n relations, and Π_{n+1} be the class of complements of Σ_{n+1} relations. It is clear that Σ_1 relations are just the r. e. relations and we know that there are r. e. relations which are not recursive (and, hence with complements not in Π_0). It is also possible to show that there are Σ_1 relations not in Π_1. Generalizing these assertions to any positive n, we arrive at the arithmetical hierarchy.

6.8.6. A relation is Σ_{n+1} iff it can be expressed as $\exists x R$, where R is Π_n; it is Π_{n+1} if it is the complement of a Σ_{n+1} relation. A relation is arithmetical if there is some n such that it is Σ_n (or Π_n). When $m > n$, there are always Σ_m and Π_m relations which are neither Σ_n nor Π_n. For each positive n, there are Σ_n relations which are not Π_n and Π_n relations not Σ_n.

An additional link is introduced by taking $\Pi_n \cap \Sigma_n$ as Δ_n for all positive n. We shall leave out some of the nicer properties of Δ_n except to mention:

6.8.7. There are Δ_{n+1} relations which are neither Σ_n nor Π_n, and there are Σ_{n+1} (and Π_{n+1}) relations which are not Δ_{n+1}. Therefore, we get a convenient hierarchy for classifying arithmetical sets.

The analytical hierarchy is introduced by using quantifiers over sets of natural numbers. Once set quantifiers are present, it is generally easy to get rid of number quantifiers. Hence, the presence of set quantifiers makes bigger increases in the complexity of the definitions. The usual notation is to put $\Sigma_0^1 = \Pi_0^1 =$ the class of all arithmetical relations. Then Σ_{n+1}^1 is again defined as projections (but on function axes or using an existential quantifier over sets) of Π_n^1 relations, and Π_{n+1}^1 relations are again the complements of Σ_{n+1}^1 relations. $\Delta_n^1 = \Sigma_n^1 \cap \Pi_n^1$. Hence, Δ_0^1 is also the class of all arithmetical relations. We have here again a hierarchy with the same properties as described in 6.8.6 and 6.8.7 for arithmetical sets.

One interesting earlier result was that the Δ_1^1 sets are exactly the hyperarithmetical sets which are closely related to the set O of recursive ordinal notations.

When we come to further extensions of recursion theory, we find a number of different directions. One area is the study of induction definitions as an abstraction from generalizations of recursive definitions and of formal systems in the sense of recursively axiomatizable theories. For example, if we view a rule of inference $X \to x$ as the generation of a sentence (the conclusion) from a set of sentences (the premisses), then we can also think of it as a mapping f from subsets X of the set A of all sentences into single sentences x. If we define a new mapping $g(x)$ over subsets of A by the definition: $g(X) = \{x \mid (\text{for some } Y \subseteq X, \ Y \to x)\}$, then we get a mapping from subsets of A to subsets of A. We see that $g(B) \subseteq B$ then say simply that B is a subset of A closed under the rule $X \to x$.

Clearly such rules or mappings are intended to be monotone in the sense that if $X \subseteq Y \subseteq A$, then $g(X) \subseteq g(Y)$, since the consequences of a smaller set are included among the consequences of a larger set. The primary generalization of inductive definitions is to say that every monotone mapping q of subsets of A into subsets of A defines inductively the smallest set closed under the mapping, i.e., $\cap \{X \subseteq A \mid g(X) \subseteq X\}$. At first, the handling of inductive definitions was not common knowledge; for example, the set O of recursive ordinal notations can be given as a simple induction but was at first brought into the obvious final form with great difficulty[1].

There are elaborate attempts in trying to generalize recursion theory to "higher types" or to recursive functionals of sets, sets of sets, etc. of natural numbers. These are closely related to the general study of inductive definitions.

Another direction of generalization is to α-recursion theory with transfinite ordinals α. This is related to Gödel's constructible sets (see next chapter). The idea is to generalize the concepts of recursive functions to higher ordinals. One way is to look at Gödel's L_α for different values of α. When L_α has suitable closure properties, α is called an admissible ordinal and L_α (or sets like them) are called admissible sets[2].

When we consider sets of real numbers (or sets of sets of natural numbers) we have also a hierarchy that is similar to the analytical hierarchy, called the projective hierarchy which was introduced in descriptive set theory in 1925. In fact, several people noticed the analogy between the arithmetical hierarchy and the projective hierarchy. One striking example is Souslin's theorem that a set A is Borel iff A and its complements are "analytic" sets. This is the

---

1)   There is a long proof to show that $x\epsilon O$ is of the form $\forall X \exists y R(x, y, X)$ in S. C. Kleene, *Am. journal math.*, vol. 77 (1955), pp. 405—428. This same result is simplified in the natural manner to half a page in *J. symbolic logic*, vol. 23 (1958), p. 250.

2)   For surveys and references, see Chapters A7, B5, and C5 in *Handbook of math. logic*. A brief outline of the history of the area between recursion theory and set theory is given by J. E. Fenstad, *Logic colloquium 76*, ed. R. Gandy and M. Hyland, 1977. He brings out the importance of the works of Clifford Spector (1930—1961) and points out two little noticed contributions to the 1957 summer conference in logic. At that conference, predicative set theory was frequently discussed with Spector who was perhaps the most active participant throughout the long conference. There were suggestions of ways to go beyond the hyperarithmetical sets and recursive ordinals, and the relation to Gödel's constructible sets was remarked on (see *Summaries of talks at the Summer Institute for Symbolic Logic*, Cornell University, 1957, pp. 377—390). At that time, proof theory, predicative set theory, higher recursion theory, inductive definitions, constructible sets, and things like the later development of admissible ordinal all were thought of together in a rather vague way. Now that most of these concepts have undergone diversified but interrelated developments, there is the natural question whether some appealing unification might be forthcoming.

analogue of the theorem that a set is recursive iff both it and its complement are r. e. It is also the analogue of the definition or theorem that the Δ_1^1 sets are the sets which are both Σ_1^1 sets and Π_1^1 sets. In fact, the hyperarithmetical sets become the analogue of the Borel sets, both having hierarchies extending to the transfinite (the latter to the classical ω_1 and the former to the recursive ω_1).

Since descriptive set theory deals with sets of real numbers but has a new branch called effective descriptive set theory, it is not clear whether it should be classified under set theory or under recursion theory. It certainly contains a large part which is of interest to set theory. In particular, there is wide interest in the relation between projective determinacy (that is to say, the axiom of determinacy stated in Chapter 4, section 2, as limited to sets in the projective hierarchy) and the existence of large cardinals[1].

---

1) For a survey of descriptive set theory, see Chapter C8 of *Handbook of math. logic.*

7. HOW MANY POINTS ON THE LINE?

7.1 Cantor and set theory

Given the representation of points by real numbers, the question is equivalent to: how many real numbers? Another equivalent question is: how many sets of integers? Before Cantor there had been no sharp distinction of the size of different infinite sets; they are just infinite and therefore all very large. However, since Cantor's work, this has become a well-defined question[1].

Cantor's first paper on set theory was published in 1874[2]. He

---

1) This problem is discussed by Gödel in his paper ''What is Cantor's continuum problem?'' *Am. math. monthly*, vol. 54 (1947), pp. 515—525; reprinted with revisions, a supplement and a postscript in *Philosophy of mathematics*, ed. P. Benacerraf and H. Putnam, 1964, pp. 258—273. We quote his explanation at length (p. 515 in the original and p. 258 in the reprint):

''This question, of course, could arise only after the concept of ''number' had been extended to infinite sets; hence it might be doubted if this extension can be effected in a uniquely determined manner and if, therefore, the statement of the problem in the simple terms used above is justified. Closer examination, however, shows that Cantor's definition of infinite numbers really has this character of uniqueness. For whatever 'number' as applied to infinite sets may mean, we certainly want it to have the property that the number of objects belonging to some class does not change if, leaving the objects the same, one changes in any way whatsoever their properties or mutual relations (e.g., their colors or their distribution in space). From this, however, it follows at once that two sets (at least two sets of changeable objects of the space-time world) will have the same cardinal number if their elements can be brought into a one-to-one correspondence, which is Cantor's definition of equality between numbers. For if there exists such a correspondence for two sets A and B it is possible (at least theoretically) to change the properties and relations of each element of A into those of the corresponding element of B, whereby A is transformed into a set completely indistinguishable from B, hence of the same cardinal number.

2) See Georg Cantor, *Gesammelte Abhandlungen*, 1932. Page references in section 1 will be to this volume. In preparing section 1, I owe much to Professor Wang Xian-jun's manuscript *History of the development of mathematical logic*, section 2.

began to ask in 1873 the question whether there is a one-one correspondence between the reals and the integers. At first, he thought that there would be one. It was with difficulty that he proved the opposite and published his proof in this paper of 1874. Implicitly the concept of cardinality for infinite sets is contained in the use of one-one correspondence and the concept of countable acquires meaning through the proof of the existence of uncountable sets. It was only in the paper of 1878 that he introduced the concept of equal cardinality in terms of one-one correspondence (p. 119) and put the concept of cardinality at the center. The term was borrowed from J. Steiner (p. 151). The term "countable" was first explicitly used only in 1882 (p. 152).

In the paper of 1874, Cantor also announced the continuum hypothesis for the first time in the primitive form that the continuum is in class "two," namely in the one immediately after the class of sets with the same cardinality as that of the integers (p. 132). In fact, he stated it as a strong conjecture: "The theorem will be made clear through an inductive procedure the presentation of which we do not here enter into closely." Ever since that time Cantor had attempted to produce such a proof without success. For example, in 1883 he again announced that he hoped soon to answer the problem positively by an exact proof (p. 192). A year later at the conclusion of his chief work (1879—1884), he once more promised to give the proof in a later continuation.

In this chief work he developed the theory of infinite cardinal and ordinal numbers to the extent that is familiar today. Besides the continuum hypothesis, Cantor was very much interested in another basic principle: namely the well ordering theorem which says that every set can be well-ordered and was later proved by Zermelo (1904) by explicitly bringing out the needed assumption known since as the axiom of choice[1]. We shall refrain from discussing the central place of this theorem in Cantor's theory, the matter of extracting axioms of set theory from Cantor's work, as well as Mirimanoff's contributions to axiomatic set theory which have been

---

1)   E. Zermelo, *Math. Ann.*, vol. 59 (1904), pp. 514—516 and vol. 65 (1908), pp. 107—128.

generally unnoticed[1].

The continuum problem is important and challenging for various reasons. It is a first natural question to ask but remains unsettled for over one-hundred years. It deals with the smallest infinite sets and infinite ordinals because it can be rephrased as asking whether there are more sets of integers than there are countable ordinals. It has stimulated much interesting work: in fact, the two most interesting and fruitful ideas in axiomatic set theory, namely constructibility and forcing, were both introduced to give partial answers to this problem. Moreover, much of contemporary research in set theory is pursued with this problem directly or indirectly in mind as an unambiguous guide and unexhausted life force.

In order to explain partial results on the continuum problem, it is necessary to consider developments which contribute to these results. On the other hand, the continuum problem provides a unifying and motivating thread for an exposition of most of the central concerns of set theory.

7.2   Finite set theory and type theory

In the preceding lecture we have considered Peano arithmetic and its predicative extension PPA as well as its impredicative extension CA. It is clear that any first order system can have both kinds of extension. Before considering the question of predicativity, we discuss first finite set theory, a fragment of set theory, which is essentially equivalent to Peano arithmetic but is formally more elegant. It is also a more natural beginning level for building up more set theory.

Let us add to the basic notation of the predicate calculus the relation symbol $\in$ for membership. We would like to say that the empty set 0 exists, that if x and y exist, the resulting set by joining y as an additional member to x (viz. $x \cup \{y\}$) exists, and that these are the only sets. It is familiar that 0 and $x \cup \{y\}$ can be introduced by definition; we shall use them without elaborating. The axioms of finite set theory F are simply:

---

1)   See *From mathematics to philosophy*, 1974, pp. 210—219.

F1. $\forall z(z \in x \equiv z \in y) \supset x = y$. (Extensionality).

F2. $\exists y(y = 0)$.

F3. $\exists z(z = x \cup \{y\})$.

F4. For every formula G of F, $\forall x G(x)$ if $G(0)$ and

$$\forall u \forall v[(G(u) \wedge G(v)) \supset G(u \cup \{v\})].$$

All the familiar axioms of set theory are derivable from the axioms, except of course the axiom of infinity since F4 is really an axiom of finitude serving the role of mathematical induction. It is possible to derive all axioms of PA from F by familiar definitions of natural numbers in set theory and, conversely if we define in PA the relation $x \in y$ by "$[y|2^x]$ is odd", then F1—F4 are also derivable from PA[1].

This system F is little used because one usually does not wish to confine attention to the finite sets. Hence, usually only the axioms which are true both for finite sets and for infinite sets are given to begin with so that nothing like F4 is permitted. It is then easy to extend to the broader system simply by adding an axiom of infinity. The use of F is more like the use of PA as the first order objects in building up type theory.

In order to set up a type theory T_ω on top of F, the usual way is to repeat the move from PA to CA: introduce different kinds of variable $x_1, y_1, \ldots$; $x_2, y_2, \ldots$; etc., replace the variables x, y, etc. by x_1, y_1, etc., modify the predicate calculus in a simple and natural manner, extend the use of $\in$ and $=$, so that $x_n = x_n$, $x_n \in y_{n+1}$, $x_1 \in y_1$, etc. are the atomic formulas. The axioms are then as follows[2]:

T1—T3. same as F1—F3.

T4. Same as F4 except $G(v)$ is replaced by $v_1 \in x_2$.

T5. $\forall z_n(z_n \in x_{n+1} \wedge z_n \in y_{n+1}) \supset x_{n+1} = y_{n+1}$.

T6. $\exists y_{n+1} \forall x_n(x_n \in y_{n+1} \equiv \phi(x_n))$, where $\phi(x_n)$ is any formula not containing y_{n+1}.

T7. Axiom of choice.

---

1) See the paper with Kenneth R. Brown, *Math. Ann.*, vol. 164 (1966), pp. 26—29.

2) Compare, for example, the formulation of type theory at the beginning of Gödel's famous paper of 1931.

The exact formulation is not essential to the present purpose. The main point is the natural idea of dividing objects into layers, in this case finite sets, sets whose members are finite sets, etc. The particular formulation given here brings out one point that is not so glaringly unreasonable with other formulations in which the first order objects are not sets (but "individuals"). The point is the mutual exclusiveness of the different layers: it is supposed to be nonsense to say that $a_n = b_k$, if a_n is an object of order n and b_k of order k, $k \neq n$. But in the present formulation it is quite possible that some set a_2 given by T6 is finite and, therefore, coextensional with a set b_1. Of course this problem is present in the other formulations too because, e.g., an empty set of order n exists for each order $n \geqslant 2$, and they are to be regarded as different. By the way, for closer classification of (say) sets of integers or sets of reals, the type theory approach remains useful.

At any rate, the more convenient common practice has been for many years to use the cumulative approach so that if $n > m$, every object of order m is also of order n. As we shall soon see, this minor point entails a good deal of notational simplification.

7.3 Axiomatization of set theory

In 1908, Zermelo and Russell proposed two seemingly very different axiom systems of set theory which gradually converged and get extended until one gets what is commonly known as the system ZF[1].

The background of Russell's system has been reviewed elsewhere[2]. The general line is as follows. In 1905, J. Richard proposed in a special context the "vicious-circle principle" which roughly says that whatever involves all of a collection must not be one of the collection. Poincaré presupposed this idea in his comments on Russell's

---

1) *Math. Ann.*, vol. 65 (1908), pp. 261—281.

2) *From math to philosophy*, Chapter III, especially pp. 103—109. Compare Gödel's 1944 paper in *The philosophy of Bertrand Russell*, ed. P. A. Schilpp.

earlier discussion and Russell accepted this, both in 1906. Definitions which do not violate the above principle are called *predicative*, the others *impredicative*. This is the sense that PPA is a predicative extension of PA, while CA is an impredicative extension. The difference is that in PPA the sets are introduced with quantifiers ranging only over numbers, while in CA quantifiers over sets which involve all of the collection of sets of numbers are employed. (See Chapter 2 for the formulation of PA, PPA, and CA).

The use of impredicative definitions can be illustrated by a slight reformulation of Cantor's diagnoal argument. Suppose a function f were to enumerate all sets C of numbers. We can then get defined a set E such that:

$$(1) \qquad \forall x[x \in E \equiv x \notin f(x)].$$

Since f is supposed to count all the sets C, there is some number x_0 such that $E = f(x_0)$. But then we have: $x \in f(x_0) \equiv x \notin f(x_0)$. The impredicative feature is implicit in (1) because, in order to define the set E, we refer to all sets, including E itself. Another application of impredicative definitions is in proving that every bounded set of real numbers has a least upper bound. Thus if we construe each real number as a set of rational numbers, the least upper bound L_k of the set K of reals is given by: $L_k = \{r| \exists X(r \in X \wedge X \in K)\}$, where L_k is included in the range of the variable X.

It is one thing to accept the negative principle, to arrive at a positive proposal that agrees with the principle requires an additional idea. And the idea is nothing but first order definability or the predicate calculus as we have considered above. If we disregard the complexities connected with the noncumulative approach, the solution amounts to a fragment of the constructible sets which Gödel introduced later on. Let us say that a set x is specifiable in a structure A with a membership relation on it, iff x is $\{y \in A | A \models \phi(y, a_1, \ldots, a_n)\}$, with $a_1, \ldots, a_n$ in A. In other words, x is first order definable over A or it is defined by a formula of set theory when the quantifiers range over A and the parameters take members of A as values. Briefly, $Sp(A)$ is the notation for the set of sets specifiable in A.

Using the greatly improved conception, we can summarize the result with the help of the familiar inductive definition:

$L_0 = 0$

$L_{\alpha+1} = \mathrm{Sp}\,(L_\alpha)$

$L_\beta =$ union of all L_α, $\alpha < \beta$, where β is a limit ordinal.

In this notation, what Russell came up with was essentially $L_{\omega+\omega}$, or rather all L_α's with $\alpha < \omega + \omega$. In fact, the finite sets in the system F amounts to L_ω, and Russell's ramified hierarchy is to take L_ω as the first order and add all the finite orders $n > 1$. Writing up the axioms for $L_{\omega+\omega}$ gets quite messy especially with the distinction between types and orders. In any case such a system (call it R_ω) is rather weak and fails to accommodate certain basic arguments in ordinary mathematical discourse.

When Russell returned to examine this ramified theory more closely in 1925, he was, more than anything else, disturbed by the fact that in his system R_ω, mathematical induction and the least upper bound theorem just mentioned cannot be proved in their general form. Actually this problem of his has an easy solution which amounts to introducing general variables at the limit orders such as $\omega + \omega$. The solution has stimulated much development in predicative mathematics[1].

The system Russell actually used in 1908 and in his joint work with Whitehead is known as the theory of types which contains beyond the ramified theory an "axiom of reducibility". The effect is to cancel out the ramification and therewith the vicious-circle principle, resulting in a system essentially as the system T_ω described in the preceding section. This theory demonstrably does not describe $L_{\omega+\omega}$ because it gives many sets not in $L_{\omega+\omega}$. So far, the only natural interpretation of this theory is to think of a lot of sets not required to exist by the axioms and satisfying much stronger axioms

---

1) The first development of this observation was made in *J. symbolic logic*, vol. 19 (1954), pp. 241—266. Since that time, a good deal more has been done by S. Feferman and K. Schütte, see for example, the references listed in Schütte, *Proof theory*, 1977 and in Chapter D4 of *Handbook of math. logic.*

as well. This is commonly known as the "rank model" which is studied closely elsewhere [1]. The model of T_ω is a fragment of this rank model, in fact $D_{\omega+\omega}$:

$D_0 = 0$

$D_{\alpha+1} = \{x \mid x \subseteq D_\alpha\}$, i.e., the power set of D_α.

$D_\beta = \bigcup_{\alpha<\beta} D_\alpha$, i.e., the union of all D_α, $\alpha < \beta$, when β is a limit ordinal.

Let us now turn to Zermelo's axioms given in 1908. This system Z can be formulated in the predicate calculus in an elegant way if we follow the later practice of identifying what Zermelo called definite properties with those expressible in this first order language. The central axioms of Z are, apart from the axioms of extensionality and choice, the following three: (a) there is an infinite set; (b) for every set, its power set exists; (c) given a set x and a property (formula of Z) $\phi(y)$, there is a subset of x consisting of all members of x which satisfy $\phi(y)$ (separation or comprehension). Although Z appears very different from T_ω, the universes of sets they describe are the same, viz. $D_{\omega+\omega}$. If we think of T_ω as a cumulative theory (which is not necessary), the only difference is that Z yields somewhat more sets because its variables range over all sets of the universe while in T_ω we have only closer and closer approximations as we increase the (finite) orders of the variables.

It is well known that both T_ω and Z are adequate for ordinary mathematical discourse. But in set theory, once they are understood in the above manner, they invite further extensions. Since we have much larger ordinals than $\omega + \omega$, why stop there? Historically the first major extension is to add the axiom of replacement to Z to obtain what is commonly known as the system ZF. It is familiar that replacement requires many more ranks than $\omega + \omega$, but in the light of recent studies, it does not go very far. In fact, the axioms of ZF can be satisfied with ranks which are, by present standards, relatively low. Since the universe of sets to be described is the union of all D_α, with α any ordinal number, "large cardinals" have

---

1)   *From math. to philosophy*, pp. 181—190.

been under intensive study for the purpose of further approximation to the intended interpretation[1].

To return to the continuum problem, we shall consider the two
definite partial results which are relative to formal systems like ZF
and exploit the fact that these are indeed formal systems in which
sets are introduced by formulas in the given formal language. The
two results are (relative) consistency by Gödel and independence
by Cohen. But conceptually it is instructive to consider first an
unsuccessful attempt by Hilbert.

7.4   Hilbert's intervention

In 1925, Hilbert proposed an outline of an argument which
purported to prove the continuum hypothesis in systems like ZF[2].
At that time Hilbert had several beliefs which have since been shown
conclusively to be mistaken and which were erroneous philosophi-
cally even relative to the state of knowledge at that time. The error
was at bottom idealism in the special form of a quasi-positivism.
It is rather common for successful scientists who at first obtain
important results by fully grasping the complexity of the real world
as it bears on his studies and then begin to believe in the feasibility of a more a priori approach to new problems.

In this case Hilbert believed that all mathematical reasoning
can in some way yet to be discovered be reduced to an elementary
part of it ("finitist" in his conception). Since the more advanced
part is merely to supplement (and extend by analogy) the more
elementary part, nothing that is true in the elementary part can be
false in the expanded region. Applied to the continuum hypothesis,
he believed it to be true if we confine ourselves to ordinals and num-
bersets in the elementary part which, according to our present
knowledge, form only a fragment of even the recursive ones. The
problem for him was to prove the special case of his general belief
of the extendibility of truth in the elementary part as applied to the
continuum hypothesis. We know now that this general belief is

---

1)   See F. R. Drake, *Set theory*, 1974 (briefly, Drake 1974).
2)   D. Hilbert, *Math. Ann.*, vol. 95 (1926), pp. 161—190.

demonstrably false and that the original continuum hypothesis is totally different from the one dealing only with recursive objects.

In one respect, however, Hilbert's approach was natural and suggestive, namely that we should build up the countable ordinals and number sets simultaneously in a hierarchy with the ordinals as indices so that more ordinals lead to more sets (or rather functions) and more functions lead to more ordinals. But this does not get us very far since the difficulty was exactly that we did not have an understandable way of grasping the many complex ways of introducing these ordinals and sets. In fact, as we shall note below, this process will get stopped quite early. Some basically new idea was necessary in order to carry out this project or even just a part of it (as it turned out, consistency rather than truth).

7.5 Constructible sets

It must have been in the summer of 1930 when Gödel began to think about the continuum problem and also heard of Hilbert's proposed solution[1]. At that time, the ramified theory mentioned above was familiar to Gödel and he also knew what are nowadays called Skolem functions. These functions are typically employed in the proof of 2.2.3 about countable models. (See Chapters 2 and 5 for related considerations.)

The Skolem functions and the basic idea of the proof of 2.2.3 are rather simple. Consider a single axiom.

(6) $\forall x \exists y R(x, y)$, with no other quantifiers.

If there is a model D for (6), there is some object, say a_1, in the model. Since (6) is true in D, there is a_2 such that $R(a_1, a_2)$ is true. Continuing thus, we get a countable subset D_c of D with members $a_1, a_2, \ldots$, not necessarily all distinct, such that (6) is true in D_c. The crucial point is that we get closure from x to y once we have a suitable countable subset of D. Alternatively, given the assumption that (6) is true in D we can introduce a function (a Skolem function) f such that, for all x in D, $R(x, f(x))$. Then we can

---

1)   This statement and related ensuing reports on Gödel's pursuit of the continuum problem are drawn from conversations with him in 1976.

take a_1, $f(a_1)$, $f(f(a_1))$, ... as the desired countable set $\{a_1, a_2, a_3, \ldots\}$.

In terms of the general approach envisaged, there are two selections to be made: the method by which the ordinals are to be obtained, the method by which sets (and, in particular, the number sets) are to be introduced at order α as each countable ordinal α is introduced. Gödel thought of the ramified theory quite early so that if α is $\beta + 1$ we introduce just $\mathrm{Sp}(L_\beta)$ to make up L_α and that if α is a limit ordinal, then L_α is just the union of all L_β, $\beta < \alpha$. The problem remains to find a method to get the ordinals. Gödel spoke of experimenting with more and more complex constructions for some extended period somewhere between 1930 and 1935.

In this connection it is of interest to mention a natural idea which was proposed in 1954 for a different purpose. It seems plausible that as we build up L_α for more and more α, L_α would contain well-orderings of the natural numbers of order types greater than α. And then we are entitled to use these larger ordinals β to go to L_β. This natural idea has come to be known as the method of "autonomous progression". For the purpose of building up the ramified hierarchy from the bottom (beginning with PA or F or $L_{\omega+\omega}$ or L_α with α the first critical ε-number, etc.), the idea is demonstrably unworkable by a theorem of Spector (1955) according to which no new ordinals appear between L_ω and L_γ where γ is the recursive ω_1 since every well-ordering of natural numbers in L_γ is of the same type as some primitive recursive one[1]. In particular, the Hilbert approach would also be stopped at order γ according to a plausible interpretation of his outline.

Gödel's new idea was to take the ordinals as given in the classical way as, say, formalized in ZF. It then becomes fairly clear that in L_{ω_1} there are no more number sets than there are countable ordinals since at each order at most a countable number of new number sets are introduced and ω times ω_1 is certainly again ω_1. This would seem to give at least a model in which CH is true. But

------

1) The proposal was made in the 1954 paper listed under footnote on page 125. The result of C. Spector is given in *J. symbolic logic,* vol. 20 (1955), pp. 151—163.

the problem is not quite so simple because it has to be verified that this is indeed a model of the axioms of set theory and, moreover, as we go beyond constructible sets of order ω_1 (and the axioms of ZF do require that), new number sets may appear. In fact, it had been known since Gödel's work of 1931 that as we go up the hierarchy of impredicatively defined sets, new number sets do appear at higher orders. Even if these difficulties are resolved, we do not have a proof of the CH because all we have would be a model in which CH and the axioms of ZF are all true. But it would be close to a proof of relative consistency because it seems most likely that the constructions can be carried out in ZF.

It is not known whether the idea of taking ordinals as given occurred in 1935 or earlier. At any rate, before autumn 1935 Gödel had formulated the constructible sets in the form familiar today, proved that the axioms of set theory (including the axiom of choice) hold for it and conjecture that the CH would also hold. The proof of the conjecture (also for the generalized CH) was not given until three years later on account of his illness.

It may be of interest first to see whether the axioms of the second order arithmetic (say CA) are satisfied by L_{ω_1}.

7.5.1 L_{ω_1} is a model of CH and the second order arithmetic including the axiom of choice.

That CH is satisfied follows from what we have just said. The axiom of choice is also clearly satisfied since we can enumerate the first order formulas for each order α and well order L_{ω_1}.

The only axiom difficult to verify is the comprehension axiom:

(7) $\exists X \forall y (y \in X \equiv \phi(y))$.

Consider a typical case when $\phi(y)$ is:

(8) $\forall Y \exists Z R(y, U, Y, Z)$.

Recall that L_{ω_1} is simply the union of all L_α with α countable and all the set variables (capital letters) range over it. Of course, by the way the constructible sets are built up, there is, by definition, a set X in L_{ω_1+1} satisfying (7) for each formula $\phi(y)$. The problem is rather to find some L_α, $\alpha < \omega_1$, such that $\phi(y)$ is true when we let all the variables range over L_α, iff $\phi(y)$ is true when they range over L_{ω_1}. Once this is proved, we have found a set X in $L_{\alpha+1}$ which satisfies (7) in L_{ω_1}. Hence, the problem reduces to a lemma that is implicit in Gödel's paper of 1939 but is known in the literature as the reflection principle when dealing with the rank hierarchy. For the special case when $\phi(y)$ is (8), we have:

7.5.2 *Lemma.*

For each U in L_β, $\beta < \omega_1$, there exists $\alpha < \omega_1$ such that

$$(9) \quad \forall y \exists Z R(y, U, Y, Z) \equiv \forall y \in L_\alpha \exists Z \in L_\alpha R(y, U, Y, Z).$$

We shall omit the details of the proof of this lemma[1] except to remark that the crucial point is to use the Skolem functions so that beginning with any U in L_{ω_1} (and therefore in some L_β, $\beta < \omega_1$), we can find a large enough $\alpha < \omega_1$ so that the examples and counterexamples necessary to determine the truth of the left-hand side are obtained by the Skolem functions and both sides of (9) are true or false together for every number y.

It remains to extend the above result from CA to ZF and to make it into a relative consistency proof. In the process it will become clear that a homogeneous treatment will take care of the GCH as well.

7.6 Consistency of GCH[2]

The GCH says that, for all α, $2^{\omega_\alpha} = \omega_{\alpha+1}$. We would like to show that this is true in L. We shall outline the steps given by

---

1) See Gödel 1939, i.e., the paper in *Proc. Nat. Acad. Sci. U.S.A.*, vol. 25 (1939), pp. 220—224; or Drake 1974, p. 130 and pp. 99—100.

2) There are many different presentations of Gödel's proof in the literature. Often results implicit in the original proof of 1939 dealt with for the special purpose on hand have been extracted and sometimes generalized to receive separate treatments under different headings. For example, Drake 1974, p. 43, pp. 84—88, 99—100, 103—106, etc. This is the standard practice in mathematics and probably makes proofs easier to follow even though it may be less attractive for the purpose of grasping different parts of a proof as forming an organic whole. Perhaps it would be good to study Gödel 1939 while using some other presentations as auxiliaries to get straight on some of the details.

Gödel in 1939.

Let x be a constructible subset of L_{ω_α}, the first problem is to prove:

7.6.1 The set x belongs to $L_{\omega_{\alpha+1}}$ or, if all members of a constructible set x have orders $<\omega_\alpha$, then x has order $<\omega_{\alpha+1}$.

Once this is proved, it is clear that GCH holds in L because there are only $\omega_{\alpha+1}$ sets of order $<\omega_{\alpha+1}$ so that the totality of constructible subsets of $_\alpha$ has cardinality $\omega_{\alpha+1}$. Of course the axiom of choice AC is true in L. In addition, L is also a model of ZF because if we take a sufficiently large cardinal β (say the first inaccessible number), then L_β would be a model of ZF for the following reason. All the crucial axioms are roughly of the form (7), or just $\exists y \, \forall x \, (x \in y \equiv \phi(x))$. In each case if we take L_β as the universe, there are sets y in $L_{\beta+1}$ (by its definition) satisfying the axioms. Since, however, β is sufficiently large so that these constructible sets are subsets of some L_{ω_α}, $\omega_\alpha < \omega_{\alpha+1} < \beta$. Hence, by 7.6.1, these sets y are in L_β.

In order to prove 7.6.1, Gödel defines a closure K of L_{ω_α} by adding x which contains all sets which are needed to satisfy the definition of x and the definitions of the necessary additional sets (known as the Skolem hull)[1]. It is easily seen that the cardinality of K remains to be ω_α. He then proves a collapsing lemma:

---

[1] The inclusion of Skolem functions in Gödel's definition here uses metamathematics according to Gödel who said that the less intuitive proof in his monograph of 1940 (with additional notes for the second printing in 1951 and the seventh printing in 1966) is given to avoid the use of metamathematics. Gödel told me in June, 1977 the following formulation regarding his work in the GCH. His main achievement, he says, really is that he first introduced the concept of constructible sets into set theory defining it as in his *Proceedings* paper of 1939, proved that the axioms of set theory (including the axiom of choice) hold for it, and conjectured that the continuum hypothesis also will hold. He told these things to von Neumann during his stay at Princeton in 1935. The discovery of the proof of this conjecture on the basis of his definition is not too difficult. Gödel gave the proof (also for the GCH) not until three years later because he had fallen ill in the meantime. This proof was using a submodel of the constructible sets in the lowest case countable, similar to the one commonly given today.

7.6.2 There exists a one-one mapping f of K on M_η, where $\eta <$ $\omega_{\alpha+1}$ is the order type of the set of orders of members of K such that, for all x and y in K, $x \in y \equiv f(x) \in f(y)$ and $x = f(x)$ for all x in M_{ω_α}.

The proof is by transfinite induction on the orders of the sets in K. The basic point seems to be the fact that constructible sets are so transparent that the gaps conceal nothing and that the definition of a set gets no distortion when we delete the gaps and change the orders accordingly.

With 7.6.2 and therefore 7.6.1, Gödel concludes:

7.6.3 The axioms of ZF (including AC) and GCH all hold for L (or, if $V = L$).

Another way of stating part of 7.6.3 is:

7.6.4 If $V = L$, then AC and GCH.

At this point, Gödel observes that the notion of constructible sets (and L) can be defined in ZF. Hence, 7.6.4 is a theorem of ZF. Next, Gödel goes on to sharpen these results to get a syntactical result of relative consistency. Since L can be defined in ZF, we can consider statements of ZF relativized to L, that is to say, reconstrued so that the variables range over L. We can of course prove in ZF that $L \subseteq V$. A formula ϕ of set theory is absolute (say, between L and V) iff for any assignment f from L, $V \models \phi(f)$ iff $L \models \phi(f)$. In other words, when the quantifiers are restricted to L, the relation ϕ on objects in L remains unchanged. It turns out that the definition of L in ZF is absolute in the above sense[1]. Hence, for all $x \in L$, $L \models x \in L$ iff $V \models x \in L$; in other words, $L^L = V^L$. Since all these arguments can be carried out in ZF, we have:

7.6.5 ZF $\models (V = L)^L$.

It is possible to carry out the proof of 7.6.4 relative to L so that we can also deduce $(AC)^L$ and $(GCH)^L$ from $(V = L)^L$ in ZF. Since we can also prove $(ZF)^L$ in ZF, if there were a contradiction from ZF plus $V = L$, there would be a contradiction from ZF alone. The situation is analogous to the translation of non-Euclidean geo-

---

1) For details see Gödel's monograph or Drake 1974. The latter makes a separate general development of absoluteness.

metries into the Euclidean so that any contradiction in the former would yield one in the latter.

7.7 Constructibility

We shall mention some properties of constructible sets without giving any proofs or even definitions of the concepts involved. The interested reader can look up the references given.

In the initial announcement, Gödel mentioned some consequences of the axiom of constructibility in descriptive set theory[1].

7.7.1 In L, there is a Δ_2^1 (PCA $\cap$ CPCA) set of real numbers which is not Lebesgue measurable and does not have the Baire property, there is a CA set which does not have the perfect subset property.

7.7.2 In L, there is a PCA well-ordering of the real numbers.

From 1960 on, many more results have been proved about constructible sets. First, there are large cardinals which imply there are nonconstructible sets and that the constructible universe is small. Large cardinals also yield results quite different from 7.7.1.

7.7.3 If there is a cardinal α such that $\alpha \rightarrow (\omega_1)^{<\omega}$, then every set definable in L is countable, there is a Δ_3^1 nonconstructible subset of ω (in particular, a set $O^{\#}$ which is essentially a truth definition of L), and there is a proper class (of ordinals) of discernibles for L.

7.7.4 If there is an (uncountable) measurable cardinal, then every PCA set is Lebesgue measurable, has the Baire property, and has the perfect subset property[2].

In recent years a detailed study of L has been made by Ronald Jensen and others. For example, the negation of Souslin's hypothesis is derived from $V = L$, combinatorial consequences are introdu-

---

1) See *Proc. Nat. Acad. Sci. U.S.A.*, vol. 24 (1938), pp. 556—557, and the 1951 notes added to the 1940 monograph. Further results on projective sets in L are in J. W. Addison, *Fund. math.*, vol. 46 (1959), pp. 123—135, 337—357.

2) For 7.7.3 see Drake 1974, Chapter 8. For 7.7.4, Robert Solovay, in *Foundations of mathematics*, 1969, pp. 58—73. By the way, it is also known that all Σ_2^1 or Π_2^1 subsets of ω are constructible, see Drake 1974, p. 164.

ced (for example, the diamond and square principles), theories of fine structure, morasses, and mice are introduced[1].

7.8 The continuum problem

Gödel continued to work in mathematical logic after publishing his results on the GCH in 1939 (his paper) and 1940 (his monograph). He obtained results in set theory which remain unpublished today. In 1941 he obtained a general consistency proof of the axiom of choice by metamathematical considerations. He believed it highly likely that the proof, unlike the one from constructible sets, goes through even when large cardinals (such as measurable cardinals) are present.

In 1943 Gödel arrived at a proof of the independence of the axiom of choice in the framework of (finite) type theory. According to Gödel, the proof uses intensional considerations, the interpretation of the logical connectives changes and a special topology has to be chosen.

The method looked promising toward getting also the independence of CH. But Gödel developed a distaste for the work and did not enjoy continuing it. In the first place, it seemed at that time he could do everything in twenty different ways and it was not visible which was better. In the second place, he was at that time more interested in philosophy. In 1977 he expressed regret that he had not continued the work. If he had continued with it, he would probably have got the independence of CH by 1950, and the development of set theory would have progressed faster.

He was interested in the philosophy of Leibniz, particularly the universal characteristic, and in the relation between Kant's philosophy and relativity theory. The latter interest led him to work on the general theory of relativity from 1947 to 1950 or 1951.

Gödel wrote a philosophical paper[2] on the continuum problem

---

1) See various expositions by K. J. Devlin: (1) Chapter 5B in *Handbook of math. logic*, (2) *Aspects of constructibility*, 1973; (3) with H. Johnsbraten, *The Souslin problem*, 1974; (4) *The axiom of constructibility*, 1977. It is announced in (1) that a revised edition of (2) is under preparation.

2) This is the paper listed under footnote 1) on page 119.

in 1947 and probably did not revive his interest in the problem until 1963 when he studied Paul Cohen's proof of independence and communicated it for publication. He did work on the continuum problem off and on since 1963. For example, in 1975 or 1976 he said he was completing the manuscript of a mathematical paper on the continuum problem.

The paper of 1947 explains the iterative concept of set and argues for the meaningfulness of the continuum problem on several levels. Since (parts of) the iterative concept have been formulated so precisely in systems like ZF, the continuum problem undeniably retains at least this meaning: to find out whether an answer, and if so, which answer, can be derived from these axioms of set theory. Gödel goes on to say that a proof of the undecidability from the accepted axioms would by no means solve the problem, because "the set-theoretical concepts and theorems describe some well-determined reality, in which Cantor's conjecture must be either true or false. Hence its undecidability from the axioms being assumed today can only mean that these axioms do not contain a complete description of that reality."

Gödel gave two reasons for conjecturing that CH is undecidable in systems like ZF. "The first results from the fact that there are two quite differently defined classes of objects both of which satisfy all axioms of set theory that have been set up so far." (A suggestion is made to the effect that from an axiom stating some maximum property of the system of sets, whereas $V = L$ states a minimum property, the negation of Cantor's conjecture could perhaps be derived.) "A second argument in favor of the unsolvability of the continuum problem on the basis of the usual axioms can be based on certain facts (not known in Cantor's time) which seem to indicate that Cantor's conjecture will turn out to be wrong."

As we know, the conjecture of undecidability was verified in 1963. But the conjecture of CH being false remains open and there is some doubt as to whether Gödel continued to believe in this conjecture during the last few years of his life.

7.9 Set theory since 1960

The study of large cardinals began to yield surprising results around 1960. For example, it was found that the first inaccessible cardinal (considered large previously) is much smaller than the first measurable cardinal. Work in infinite combinatorics[1] with various model-theoretic consequences began to introduce ramifications and refinements in the spectrum of large cardinals. The measurable cardinals were seen to imply the existence of nonconstructible sets[2].

In 1963 a manuscript of Paul J. Cohen began to circulate in which the method of forcing was introduced[3]. The method not only proves the independence of CH from the usual axioms of set theory but is very flexible in introducing many models to yield independence proofs for many mathematical statements (for example, Souslin's hypothesis, various forms of the axiom of choice, Kurepa's conjecture, etc.). Cohen's method gives a way of expanding models by adding new sets in a very economical manner. The new models are made to have only properties forced by the addition of the new sets; the interaction of the new sets with the original models is kept at the necessary minimum. As a result, one is able to show that the expanded domain still satisfies the axioms and the new sets are able to serve the original purpose of negating a statement to show its independence without confusing wide ramifications.

The method of forcing uses initially the ramified hierarchy familiar from the constructible sets. A natural question to ask from a naive point of view is whether one can also do it with an unramified (the rank) hierarchy. I remember trying to do this in 1965, but I did not get a clearcut answer. In the autumn of 1965, Solovay

---

1) For a recent survey, see Chapter B3 in *Handbook of math. logic.*

2) This result was due to D. S. Scott, *Bull. Acad. Polon. Sci.*, vol. 7 (1961), pp. 145—149. Measurable cardinals were first considered by S. Ulam, *Fund. math.*, vol. 16 (1930), pp. 140—150. The area of measurable cardinals contains various elegant results which are summarized in J. Shoenfield, *Logic colloquium 69*, ed. R. D. Gandy and C. M. E. Yates, 1971, pp. 19—49, and in Drake 1974.

3) The original manuscript (*Independence of the axiom of choice*, Stanford University, 1963) contains various errors which were soon corrected. A paper in two parts followed soon, *Proc. Nat. Acad. Sci. U.S.A.*, vol. 50 (1963), pp. 1143—1148 and vol. 51 (1964), pp. 105—110 (communicated by Gödel). In the spring of 1965, Cohen went to visit Harvard and gave a course on logic, resulting in the monograph *Set theory and the continuum hypothesis*, 1966.

was using Borel sets of positive measure as forcing conditions. To summarize various calculations with the forcing conditions, he took to calling the combination of conditions forcing a statement as adding up to the "value" of that statement. He soon realized that one could also start with Boolean-valued sets at the beginning; Scott knew Solovay's earlier suggestion and came independently somewhat later to the same realization. Meanwhile Vopenka was also using Boolean-valued models[1] in 1965.

Even though the use of Boolean models got a good deal of publicity in 1966 and soon after, the general opinion now seems to think of it as a convenient variant of forcing. For example, Shoenfield gives a presentation of unramified forcing which demonstrates that the two approaches come to the same thing[2]. When we use Boolean-valued models, it is highly plausible that for any complete Boolean-algebra B, the axioms of ZF are true for the B-valued sets. And there is in fact a general theorem to this effect. The corresponding result for forcing is also true but does not appear equally direct at first sight. However, when it comes to choosing a particular Boolean algebra or a particular set of forcing conditions to prove a particular statement independent, most set theorists seem to find it psychologically more helpful to think in terms of forcing.

Boolean-valued models of first order theories have been around for a long time but they had not been employed for independence proofs until after the discovery of forcing. In the study of models of intuitionistic logic, considerations similar to forcing had also been employed before 1963, but again the similarity was only noticed afterwards[3].

---

1)   P. Vopenka, *Bull. Acad. Polon. Sci.*, vol. 13 (1965), pp. 189—192. For an account of the history of Booleanvalued models, see J. L. Bell, *Boolean-valued models*, 1977, pp. xi—xviii.

2)   There was a summer institute devoted to axiomatic set theory in 1967. The proceedings have been published in two volumes by Am. Math. Soc.: *Axiomatic set theory*, vol. I, 1971, ed. D. S. Scott, vol. II, 1974, ed. T. J. Jech. Shoenfield's paper appears in vol. I, pp. 331—355.

3)   Traditional work on Boolean-valued models is reported in H. Rasiowa and R. Sikorski, *The mathematics of metamathe-matics*, 1963. The analogy between forcing and models of intuitionistic logic is examined in M. C. Fitting, *Intuitionistic logic, model thoery and forcing*, 1969.

7.10 GCH and the relativity of cardinality

One striking consequence of the method of forcing is the realization that cardinal numbers are extremely relative as we move from models to models of set theory. Constructivity shows that by cutting down on the universe of sets, we get more or less a unique model in which the cardinals are well-behaved in a determinate manner so that, for example, the GCH is true in the resulting more or less minimum model. Forcing shows that there are unlimited possibilities of expanding simple models to get new models to satisfy almost any strange requirements on the behavior of cardinals. For example, one could get models of set theory in which 2^ω is ω_{21} and 2^{ω_1} is ω_{502}, etc.

In fact, there has been little progress so far on unambiguous answers to the GCH. A great deal of work has been devoted to the determination as to what possible values 2^{ω_α} could have, especially when ω_α is a singular cardinal. The cofinality of a cardinal α, $cf(\alpha)$, is the smallest cardinal β such that α is the union of β cardinals $< \alpha$. A cardinal number α is regular if $cf(\alpha) = \alpha$, otherwise singular (i.e., $cf(\alpha) < \alpha$). For example, $\omega, \omega_1, \omega_2$ are regular, ω_ω is singular since $cf(\omega_\omega) = \omega$.

We proceed to give a brief summary of results on CH and GCH[1]. A general way of looking at GCH is to ask the possible values of a function $f: \omega_\alpha \to 2^{\omega_\alpha}$, or $f(\omega_\alpha) = 2^{\omega_\alpha}$. The only classical limitation (apart from obvious ones such as f must be monotone increasing, etc.) is[2]:

7.10.1 $f(\omega_\alpha)$ cannot have cofinality $\leqslant \omega_\alpha$.

7.10.2 If ω_α is singular and $f(\beta) = \gamma$ for all sufficiently large $\beta < \omega_\alpha$, then $f(\omega_\alpha) = \gamma$.

---

1) For a recent review, see D. A. Martin's paper in *Math. developments arising from Hilbert problems*, 1976, pp. 81—92. It is noted there that the independence proof (of CH) has also positive consequences. If A is a statement about integers and reals only provable from ZFC $+$ CH, then A is provable from ZFC. If A is about integers only and provable from ZFC $+ \neg$ CH, then A is provable in ZFC.

2) J. König, *Math. Ann.*, vol. 60 (1905), pp. 177—180, 462. Only 7.10.1 is due to König, 7.10.2 is not attributed to anybody specially.

In particular, $f(\omega)$ cannot be ω_ω. Cohen and Solovay showed that 7.10.1 is the only restriction on the value of $f(\omega)$. Easton extends the result to GCH[1]:

7.10.3   For all regular ω_α, $f(\omega_\alpha)$ can take any values not excluded by 7.10.1.

Since 1964, the question of determining the values of $f(\omega_\alpha)$ for singular ω_α has been known under the name of singular cardinals problem and stimulated much research. A simpler form of the problem is whether GCH can hold below a singular cardinal α but fail at α. After a long pause, Silver obtained results which are contrary to previous expectations[2]. Examples of his results are:

7.10.4   If ω_α is singular and $cf(\omega_d) > \omega$, and if $f(\omega_\beta) = \omega_{\beta+1}$ for all $\beta < \alpha$, then $f(\omega_\alpha) = \omega_{\alpha+1}$.

7.10.5   If GCH holds for every singular ω_α with $cf(\omega_\alpha) = \omega$, then GCH holds for all singular cardinals.

Shortly afterwards, Jensen proved, in effect[3]:

7.10.6   If $0^\#$ does not exist, then the essential structure of cardinalities and cofinalities in L is retained in V. Briefly, if GCH fails at a singular cardinal, then $0^\#$ exists.

Generally, the problem of singular cardinals of cofinality ω remains open.

In Gödel's paper of 1947, it was suggested that new axioms of infinity might help to decide CH. So far there has been little progress in this direction. In fact, Cohen's method seems to suggest the opposite expectation[4]. Let A be a strong axiom of infinity asserting the existence of a large cardinal α. A mild Cohen extension of a model M relative to α uses a partially ordered set P in M such that $|P| < \alpha$ is true in M. All the standard properties of α are preserved in such extensions. However, it is generally possible to make such extensions in which CH can be true or false. In other words, there generally are models of ZFC + A + CH and of ZFC +

---

1)   W. Easton, *Ann. math. logic*, vol. 1 (1970), pp. 137—178.

2)   Jack Silver, *Proc. Int. Cong. Math.* 74, 1975, pp. 265—268.

3)   K. I. Devlin and R. B. Jensen, *Kiel logic colloquium 74*, 1975, pp. 115—142.

4)   See Martin's paper under footnote 1) on page 139.

$A + \neg$ CH, if there are models of ZFC $+ A$ at all.

Large cardinals have, nonetheless, yielded partial results on the GCH[1]:

7.10.7 If there exist compact cardinals, then GCH holds for all sufficiently large singular strong limit cardinals ω_α (i.e., singular ω_α such that for all $\beta < \omega_\alpha$, $2^\beta < \omega_\alpha$).

Another approach to the CH is by looking at simple sets of reals either to find counterexamples or to find clues for ways of treating arbitrary sets of reals. For example, it is known that Borel sets all are countable or have cardinality 2^ω. By assuming the existence of measurable cardinals, Solovay shows that every PCA set has the same property, and Martin shows that every PCPCA set has cardinality ω, ω_1, ω_2 or 2^ω. Another approach is to use prewellorderings and the axiom of determinacy as applied to projective sets (called PD). Larger and larger bounds are found for the length of projective prewellorderings as we go up to the projective hierarchy[2]: "Thus PD extends the pattern derived from MC (which extends that derived in ZFC alone) and reinforces the suggestion that the answer to CH may be negative."

7.11 The method of forcing

We return to a general discussion of the ideas involved in the method of forcing. Since there are so many presentations and ramifications in the literature[3], we shall try to avoid technical details as much as possible. Moreover, we shall consider only a simple case, disregarding alternative approaches and freely assuming restrictions (such as beginning with a countable model) which are convenient but not necessary.

One possible idea is this. Let M be a given countable model

1) R. M. Solovay, *AMS Proc. pure math.*, vol. 15 (1974), pp. 365—372.

2) Martin, op. cit., p. 90.

3) To begin with there are Cohen's paper and book listed under footnote 3) on page 137. A recent summary is Chapter B4 in *Handbook of math. logic.* Two monographs published in 1971 are: U. Felgner, *Models of ZF set theory* and T. J. Jech, *Lectures in set theory.* An expanded new edition of Jech's book is awaiting publication.

of ZF, which could be thought of as an initial segment of L. To this model we can associate a countable language Y and an enumeration of all the statements of Y. Every statement is either true or false in M. There is a standard way of adjoining a set b of natural numbers to M to get $M(b)$. Let us leave the set b unfixed and consider $M(b)$ for all subsets b of ω. Extend the language Y to Y^* by adding the new constant b in the natural manner. We can again enumerate all the statements in Y^*: ϕ_0, ϕ_1, etc. in some suitable manner. Each statement is either true or false in $M(b)$, once a fixed set b is chosen. In particular, the statements $0 \in b$, $1 \in b$, etc. and their negations all appear in the sequence ϕ_0, ϕ_1, etc.

Consider now the function $f:2^\omega \to 2^\omega$ defined as follows.

7.11.1   For each subset b of ω, $f(b)$ is the subset of ω determined by the following condition: $i \in f(b)$ iff $M(b) \models \phi_i$ (and, therefore, also $i \notin f(b)$ iff $M(b) \models \neg \phi_i$), where ϕ_i is determined by the uniform ennumeration of the statements of the same language Y^*.

This is roughly a mapping of real numbers between 0 and 1 into real numbers between 0 and 1 except that we need a different topology from the usual one (with each b corresponding to a real number as a sequence of binary digits in the obvious manner). We shall, at the risk of confusion and mistakes, talk as though we were dealing with real numbers between 0 and 1.

A first rough idea is this. Let us see how $f(b)$ changes with b. If $b \in M$, then assuming (as we can) that every member of M has a name in Y and Y^*, the function $f(b)$ must be discontinuous at b because among the statements in Y^*, we have a ϕ_j which is $n_b = b$ where n_b is the name of the set b in M. This is so because at no point near b is $n_b = b$ true again. This leads to the conclusion that if we are looking for b not in M, we have to confine our attention to points b at which $f(b)$ is continuous. We note that our first attempt at independence proofs is to find a subset b of ω such that (i) $b \notin M$; (ii) $M(b) \models \text{ZF}$. From (i) and (ii), there are standard arguments to infer that $V = L$ is independent of ZF.

Once we hit on the idea of continuity, we are led to consider finite approximations to b. If $f(b)$ is continuous at b, then the truth values of the statements do not change abruptly. Hence, for each statement ϕ, there is a neighborhood of b or a finite segment of length n of b such that ϕ has the same truth value for all reals with the same initial n digits as b. There remain two problems now. How do we know that $f(b)$ is continuous at any b (or better for a lot of b's)? What is the relation between continuity at b and the desired objective (ii), i.e., $M(b) \models \mathrm{ZF}$.

On the first question there are certain classical results which suggest that we do have a lot of b's. One result is Blumberg's theorem:

7.11.2 For every real function $f(x)$ on $\mathrm{I} = (0, 1)$, there is an everywhere dense set E such that $f(x)$ is continuous on E relative to E. A perhaps more relevant result is[1]:

7.11.3 Every Baire function is continuous "up to" sets of the first category (or, up to a meager set), i.e., continuous on a comeager set of the original space.

To apply 7.11.3 it is of course necessary to show that the function $f(b)$ defined above is a Baire function in the appropriate topological space.

By the way, I have heard and seen suggestions similar to the approach being described here several times. But I have not seen it worked out anywhere and I have not done so myself. I am merely looking at the approach more closely from my incomplete understanding of it.

Let us now turn to the second question. Roughly speaking, b must not interact with M too strongly or in some disrupting fashion in order that $M(b)$ remain a model of ZF. The frequently cited example is when b is well-ordered with an ordertype greater than the largest in the countable model M. In that case it is easy to see that $M(b)$ is no longer a model of ZF. On the other hand, given any b, the statements of the form $\bar{n} \in b$ for any numeral $\bar{n}$ (as well as their negations) must have fixed true values. But of course they change continuously with b. In fact, the truth value of each such statement is completely decided by some initial segment of b which

---

1) The result 7.11.2 is in H. Blumberg, *Trans. Am. Math. Soc.*, vol. 24 (1922), pp. 113—128. 7.11.3 is given on pp. 295—296 of Felix Hansdoff, *Set theory*, second English edition, 1962 (the German original had three editions from 1914 to 1937).

144 Popular Lectures on Mathematical Logic

defines a neighborhood. But of course, the constant b also appears in more complex statements.

We assume that B is a subset of 2^ω such that, for every b in B, the function $f(b)$ defined above is continuous on B. Say that two subsets c and d are n-close iff $i \in c \equiv i \in d$, for all $i < n$. Since continuity at b is taken to mean that the truth values of the statements (in Y^*) on $M(b)$ do not change abruptly, we have, for each b in B and each statement ϕ_i in Y^*, the following situation. For each i, there is some n such that the truth value of ϕ_i at b remains the same at all points n-close to b. If we now call all the finite segments p of b conditions, we get the following conclusion.

7.11.4 If $f(b)$ is continuous at b, then, for all i, $M(b) \models \phi_i$ iff there is some finite segment p of b such that ϕ_i is true in the neighborhood of b determined by p. Let us then say that p forces ϕ_i to be true, or in the standard notation, $p \models \phi_i$.

But 7.11.4 merely says that there exists some distribution of truth values such that $M(b) \models \phi_i$ iff there is some $p, p \Vdash \phi_i$. We have not found out what the distribution is yet. However, the mere existence of such a distribution makes plausible the now familiar proof that $M(b)$ gives a model of ZF. The proof[1] using labelling (or Gödel numbering) to represent all the terms and sentences of Y^* and $M(b)$ depends merely on the fact that we can absorb the new constant b and handle the conditions making up the set b to define all special cases of $\Vdash$ in M and verify the axioms of ZF for $M(b)$. But the definability of $\Vdash$ needs a more explicit specification of the relation $\Vdash$.

From what we have said above, it is obvious how we can deal with basic statements of the form $i \in b$ and their negations (or also all their Boolean combinations). It is when quantifiers appear that we have a problem. Since we assume that every member of $M(b)$ has a name, there is no reason why we cannot define $p \Vdash \exists x Fx$ to mean that there is a term c of Y^* such that $p \Vdash Fc$. This would not do for $\forall x Fx$ because there are infinitely many values of x so that we may need an infinite set of conditions to force all the

---

1) See, e.g., Cohen's book, pp. 113—115, 120—126.

instances of the statement. That would defeat our purpose of talking about b only by mentioning some finite segment of it each time.

Cohen uses the following definition: $p \Vdash \forall x F x$ iff there is no extension q of p, $q \Vdash \neg Fc$, for some term c. In Cohen's original treatment, he assumed sentences to be in prenex form because he treated negation in the simple form that $p \Vdash \neg \phi$ iff p does not $\Vdash \phi$. Since $\forall x$ is $\neg \exists x \neg$, a simplification has since been introduced by treating negation generally in the manner of $\forall$: $p \Vdash \neg \phi$ iff there is no extension q of p, $q \quad \phi$. We can then define $\forall$ by $\neg \exists \neg$, and forcing for $\forall$ follows as a consequence so that, in particular, there is no need to use only sentences in prenex form.

Once we settle on the definition of forcing for negation or universal quantification, the rest is as usual. For example, $p \Vdash \phi_1 \wedge \phi_2$ iff $p \Vdash \phi_1$ and $p \Vdash \phi_2$. The inductive definition thus obtained yields a simple and correct characterization of the truth values of the statements of Y^* in $M(b)$. It is then easy to show that forcing has the desired properties to be a substitute for satisfiability or truth in M (b). For each ϕ, one can then define in M the relation $p \Vdash \phi$, for arbitrary condition p, and verify the axioms of ZF, as well as $M(b) \models \neg V = L$. There is the vague idea that the "local" character of the axioms of ZF has something to do with the applicability of arguments like this. These axioms are local in the sense that every single axiom concerns only a small fragment of the universe of all sets that satisfies all the axioms. For example, an axiom giving a universal set or unlimited complement of each set would not be local in this sense. But I do not know whether this vague idea contains anything definite and of interest.

We have tried to state rather inconclusive impressions to motivate forcing. Usually one learns a new method by studying exact presentations. We proceed to give some more exact descriptions, beginning with a simple case having to do with arithmetical sets (or rather first order definable sets of the universe of hereditarily finite sets)[1].

---

1) This way of first introducing a simple case of forcing is used by Ronald B. Jensen in his monograph, *Modelle der Mengenlehre*, 1967.

7.12   Forcing in a simple setting

The discovery of irrational numbers was an important step taken in ancient times. In the nineteenth century, different ways of proving the existence of transcendental (i.e., nonalgebraic) numbers were introduced by Cantor with the diagonal argument and by others with more specific proofs establishing the transcendence of e and π.

In this century, we have proofs to give functions which are not primitive recursive, sets which are not general recursive or not recursively enumerable, or not arithmetical, not hyperarithmetical, not analytic, not projective, etc.

The simplest case to look at in connection with independence results in set theory is to find a set which is not constructible but satisfies the usual axioms of set theory when adjoined to a suitable model of these axioms. In order to motivate the method introduced by Cohen, we follow Jensen by considering first the simpler question of finding a general way of getting sets which are not arithmetical.

Since the totality of arithmetical sets is countable and there are uncountably many sets of integers, there are of course many nonarithmetical sets. In fact, there are familiar ways of getting many of them such as using stronger axioms of comprehension. But our purpose here is merely to introduce forcing in a simplified context.

In general terms, one can summarize the notion of forcing in the following manner. We begin with a language Y and a structure M of it . A new symbol G is added to Y to yield Y^* for the generic set to be introduced. The addition of G into Y induces an extended structure $M(G)$ consisting of objects obtainable from G by certain operations. The goal is to find G with suitable properties. For example, if M is a countable model of ZF, we would like to find a new set G of integers such that $M(G)$ is also a model of ZF, thereby showing that the axiom of constructibility is independent.

We assume that Y^* contains means to denote the objects of $M(G)$. But since we do not know what G is and G is not in M, we have no direct means of talking about G in Y. However, it is usually possible to talk about finite segments of G in Y.

The definition of forcing is given inductively for sentences in

Y^*. A sentence ϕ of Y^* is forced (to be true) by a finite amount of information p about G, if the truth of ϕ in $M(G)$ can be established on the basis of p and will remain undisturbed however p is extended (to any larger finite amount of information about G). We cannot, of course, expect that for any G and ϕ of Y^*, the truth of ϕ or $\neg \phi$ in $M(G)$ can be established in this way. Any set G such that the truth of every ϕ (or $\neg\phi$) in $M(G)$ can be established in this way is a generic set. Indeed, it seems difficult to see intuitively whether generic sets exist. It turns out that there are plenty of generic sets.

Let Y be the first order language of set theory and M be $\langle V_\omega, \varepsilon \rangle$, the structure of hereditarily finite sets. Let P be the partially ordered collection of conditions p where a condition p is a mapping of a finite segment of ω into $\{0, 1\}$. Intuitively, when n is in the domain of p, $p(n) = 1$ or 0 according as n belongs to G or not.

Hence, P may be thought of as consisting of all the finite paths in the full binary tree. A condition q extends $p(q \geqslant p)$ if q contains p. A particular real number or set of integers is to be represented by an appropriate sequence of conditions (viz. which are "compatible" in the obvious sense). The forcing relation is usually denoted by $\Vdash$.

Definition of forcing by induction on sentences in Y^*:

(1) $p \Vdash x \in y$ iff $x \in y$

(2) $p \Vdash x = y$ iff $x = y$

(3) $p \Vdash G(x)$ iff $p(x) = 1$

(4) $p \Vdash \phi \wedge \psi$ iff $p \Vdash \phi$ and $p \Vdash \psi$

(5) $p \Vdash \exists x \phi(x)$ iff $\exists x\, (p \Vdash \phi(x))$

(6) $p \Vdash \neg \phi$ iff   For all q, if $q \geqslant p$, $\neg(q \Vdash \phi)$.

The definition differs from the familiar truth definition only in clause (6). If we were using a full (infinite) set G of integers instead of its infinitely many finite segments we would say G makes $\neg\phi$ true iff G does not make ϕ true. As it is, we cannot reasonably say:

$$p \Vdash \neg\phi \quad \text{iff} \quad \neg(p \Vdash \phi).$$

Since p contains only a finite part of G, it is possible that a larger part q of G might make ϕ true, and we would then have both ϕ and $\neg\phi$ forced to be true by the set G yet to be secured.

It is fairly easy to prove the following theorems:

(a) $\neg(p \Vdash \phi \land p \Vdash \neg\phi)$.

By clause (6), since $p \geqslant p$.

(b)   If $p \Vdash \phi$ and $q \geqslant p$, then $q \Vdash \phi$.

Consider all six clauses (1)—(6) one-by-one.

(c)   Either $p \Vdash \phi$, or $\exists q$, $q \geqslant p$ and $q \Vdash \neg\phi$.

Immediate by (6). It is a remarkable property of forcing.

(d) $p \Vdash \phi \leftrightarrow \forall q$, if $q \geqslant p$, then $\neg(q \Vdash \neg\phi)$.

We write $p < G$ whenever p is a segment of G, or for all n, in the domain of p, $p(n) = 1$ iff n belongs to G. By the way, when $x \notin \omega$, every $p \Vdash \neg G(x)$, by (6) and (3).

There are different ways of defining generic sets.

D1.   *A set G is generic iff for every sentence $\phi(G)$ of Y^*,*

$$M(G) \models \phi(G) \text{ iff } \exists p(p < G \text{ and } p \Vdash \phi(G)).$$

It is natural to write $G \Vdash \phi$ instead of $\exists\, p(p < G$ and $p \Vdash \phi)$.

D2.   For every ϕ in Y^*, $G \Vdash \phi$ or $G \Vdash \neg \phi$.

The crucial matter is to prove that generic sets exist.

Theorem.   For any q, there is a generic set G, $q < G$.

Since Y^* is a countable language, we can enumerate all its sentences: ϕ_1, ϕ_2, etc. It is an easy corollary of (6) or (c) that for every q and every ϕ_i, there is an extension p such that $p \Vdash \phi_i$ or $p \Vdash \neg \phi_i$. Thus, either q has an extension p, $p \Vdash \phi_i$ or q has no such extension; but in the second case, by (6), $q \Vdash \neg\phi_i$ and, therefore, every extension of q forces $\neg\phi_i$.

We can, therefore, define G simply as p_0, p_1, etc. as follows.

$p_0 = q$

p_{i+1} = the smallest extension of p_i such that $p_{i+1} \Vdash \phi_i$ or
      $p_{i+1} \Vdash \neg \phi_i$.

Hence, G is generic in the sense of D2, i.e., for every ϕ in Y^*, $G \Vdash \phi$ or $G \Vdash \neg \phi$.

To see that G is also generic by D1, we apply induction on the complexity of ϕ. The crucial case is negation. By induction hypothesis,

$M(G) \models \phi$ iff $G \Vdash \phi$.

Therefore, by D2, $M(G) \models \neg\phi$ iff $\neg(M(G) \models \phi)$ iff $\neg(G \Vdash \phi)$ iff $G \Vdash \neg\phi$.

It seems surprising that we can so painlessly find so many generic sets. Roughly speaking, there are uncountably many sets of integers, a countable language leaves much that is indeterminate. This becomes specially perspicuous when we deal with a countable structure on the language. The special property of a generic set G is that the truth of every sentence ϕ in $M[G]$ is determined already by finite segments of G. This has the consequence that G must be very different from the arithmetical sets which are definable in M.

Theorem. Every arithmetical subset of a generic set is finite.

Let A be an arithmetic subset of G, such that $\phi(x)$ defines A. Since $A \subseteq G$, there is $p < G$, $p \Vdash A \subseteq G$.

$p \Vdash \forall x(\phi(x) \rightarrow x \in G)$.

For all x, if $p \Vdash \phi(x)$, then $p \Vdash x \in G$.

But then, for all x, if $\phi(x)$, then $p(x) = 1$.

Since p is a finite condition, $\phi(x)$ is satisfied for only finitely many x, and A is a finite set. Hence, G cannot be arithmetical by the following lemma since $G \subseteq G$.

Lemma. No generic G is finite.

A set G is infinite if $\forall x \exists y \ (x \in y \text{ and } y \in G)$.

Hence, if G is finite, then there is some condition p which forces some x to be an upper bound of G. This is not possible because of (a) and (c) for all conditions which are true before a generic set is chosen. By (a) and (c) it would follow that no extension of p can falsify the statement that x is an upper bound. But of course, we can extend p to some q so that q forces more members into G.

The point is that G retains the crucial properties of p in the matter of forcing statements in Y^*. In particular, if there is an extension of p forcing some ϕ and p is in G, there is also an extension of p in G to force ϕ.

For the same reason, every p forces G to have infinitely many primes, to be nonconstructible, etc.

7.13   Nonconstructible sets

We shall sketch a proof of the independence of $V = L$ from ZF (plus AC and GCH) in a manner that is familiar today[1].

By the Löwenheim-Skolem submodel theorem, since L is a model of ZF, it has a countable elementary submodel of ZF. By "collapsing" down, we get a segment L_α, α countable, of L which is a model of ZF. Take $M = L_\alpha$ as the given countable model. The ordinal α is said to be the ordinal of M. We now adjoin a subset G of ω to M (for the definition of Sp see section 3 of this chapter):

7.13.1   Let $M(G) = L_\beta$, for $\beta \leqslant \omega$, $M_{\omega+1}(G) = M_\omega(G) \cup Sp$ $(L_\omega \cup \{G\})$, $M_{\beta+1}(G) = M_\beta(G) \cup Sp(M_\beta(G))$, for $\beta > \omega$, and for β a limit ordinal, taken union of $M_\gamma(G)$, $\gamma < \beta$. Recalling that the ordinal of M is α, we define $M(G) = M_\alpha(G)$.

If we had begun with L, we could have defined $L(G)$ from L in the same way as we have defined $M(G)$ from L_α, using all ordinals rather than only those $<\alpha$. It is a familiar fact that $L(G)$ remains a model of ZF. The situation is, however, different with M. For example, since α is countable, we can find a well-ordering $<$ of ω of order type $\beta \geqslant \alpha$ and a set G such that $2^m 3^n \in G$ iff $m < n$. The ordinals of $M(G)$ are just those of M, but it is a theorem of ZF that the order type of every well-ordering in ZF is an ordinal in ZF. Hence, for such a G, $M(G)$ is not a model of ZF.

Even though the subsets of ω correspond roughly to the reals between 0 and 1, there is the familiar complication that different sequences of binary digits may represent the same real number. Rather the set 2^ω can be represented more directly by points in the Cantor discontinuum with the triadic notation, viz. the set of numbers t, $t = t_1/3 + t_2/9 + t_3/27 + \ldots$, with $t_n = 0$ or 2.

------

1)   The sketch follows lecture notes by D. A. Martin prepared in the spring of 1967. Many similar treatments have appeared in the literature; for example, A. Mostowski, *Constructible sets with applications*, 1969; G. Takeuti and W. M. Zaring, *Axiomatic set theory*, 1973. Many people came independently to proofs along the same line. Often it is said that R. M. Solovay used a proof of this type first.

The accepted terminology is to say: give $\{0, 1\}$ the discrete topology and the set $\{g | g : \omega \to \{0, 1\}\}$ the product topology[1]. This gives us a topology on 2^ω. We may think of the concept of neighborhood in terms of the notion of two subsets b and c of ω being n-close for some n, as mentioned before. A basic open set is essentially the set of all G which satisfy some condition p, or a finite set of sentences of the form $m_i \in G$ and $\neg \, n_j \in G$.

7.13.2   The open sets are the countable unions of basic open sets, since there are only countably many basic open sets. A subset B of 2^ω is dense iff it has a nonempty intersection with every nonempty open set. B is nowhere dense iff its complement contains a dense open set. B is meager (first category) iff it is the union of countably many nowhere dense sets. B is comeager iff the complement of B is meager. B is a Borel set iff it belongs to the σ-field of subsets of 2^ω generated by the open sets.

Some standard theorems on these standard concepts in topology are relevant:

7.13.3   Every comeager set is dense; an open meager set is empty. (Baire category theorem.)

7.13.4   Every Borel set has the Baire property, i.e., can be approximated by an open set up to a difference of some meager set.

As before, let Y be the language of M and Y^* be that of $M(G)$. We assume that these include the ramified language for all indices $\beta < \alpha$, α being the ordinal of M and $M(G)$. First we give the ordinary truth definition in terms of the ramified hierarchy for Y^* and then move to "unlimited" statements. For example, the unlimited statement $\exists \, x F x$ is true iff there is some term c such that Fc is true, only now c can be of any index $\beta < \alpha$.

The inductive definition of truth omitted here makes the following theorem clear:

7.13.5   For every ϕ in Y^*, $T(\phi) = \{G | \phi \text{ is true in } M(G)\}$ is a Borel set.

---

1)   See, e.g., K. Kuratowski, *Introduction to set theory and topology*, second English edition, 1972, pp. 210—216.

This is so essentially because for basic sentences ϕ, $T(\phi)$ is an open set and the inductive clauses only take complements and (at most countable) unions and intersections.

We now make a parallel definition of forcing for all statements of Y^*. The only crucial difference comes in when we pass from ϕ to $\neg\phi$. In the truth definition, we take the complement of $T(\phi)$ as $T(\neg\phi)$. In the forcing definition, we take the interior of the complement of $F(\phi)$ as $F(\neg\phi)$, where $F(\phi)$ is the set of G's such that there is a condition p (a finite segment of G), $p \Vdash \phi$. Hence, in each case, the difference is but by a meager set. Since the union of countably many meager sets remains meager and since there are only countably many statements in Y^*, we get:

7.13.6   For each ϕ, $T(\phi)$ differs from $F(\phi)$ by a meager set; moreover, there is a comeager subset B of 2^ω such that every G in B has the property: for every ϕ of Y^*, $G \in T(\phi)$ iff $G \in F(\phi)$. Any such subset G of ω is a generic set in the sense that for any such G, forcing gives exactly the set of true statements as being forced (to be true).

Let us use again the notation $G \Vdash \phi$ to mean there is some condition p contained in G, $p \Vdash \phi$. We can restate the property of generic sets as a definition.

7.13.7   A set G is generic over M iff for every ϕ in Y^*, $G \Vdash \phi$ iff $M(G) \models \phi$.

From this definition it follows that:

7.13.8   A set G is generic over M iff for every ϕ in Y^*, $G \Vdash \phi$ or $G \Vdash \neg\phi$.

7.13.9   For each ϕ, $F(\phi) \cup F(\neg\phi)$ is dense and open.

7.13.10   A set G is generic iff G belongs to every dense open set definable in M.

7.13.11   For every ϕ in Y^*, $T(\phi)$ is a comeager set iff every generic set G forces ϕ.

We shall omit the proofs. Since $T(V \neq L)$ is a comeager set, every generic set G makes it true in $M(G)$. We have briefly indicated before why the axioms of ZF are true in $M(G)$, we shall omit the proof of this as well as of the truth of AC and GCH in $M(G)$.

A more general approach to forcing begins with an arbitrary partially ordered set P in M, as a generalization of the set of all conditions p in the original sense. Two members p and q of P are incompatible, iff there is no member of P greater than both (generalizing the original definition that they have no common extension). A subset S of P is compatible iff any two elements of S are compatible. A subset S of P is dense in P iff every element of P is smaller than some element of S (generalizing the definition that every condition is contained in some member of S).

A subset G of P is said to be generic over M when G is compatible, every p in G has an extension in G, and every dense subset S of P (in M) intersects G. Using this concept of being generic, one can define forcing more simply:

$p \models \phi$ in Y^* iff for every G generic over M and containing p, $M(G) \models \phi$.

For these notions we have:

7.13.12 If M is countable, then every member of P is contained in some subset G of P generic over M; i.e., every condition p can be extended to get a generic set.

Since M is countable, we can enumerate all dense subsets S_1, S_2, etc. of P which belong to M. Let p_i be an element of S_i extending p_{i-1} (with $p = p_0$). Let $G = \{ p_i | i < \omega \}$.

7.14 Independence of the CH

We conclude this chapter[1] with an outline of a proof of the independence of CH from ZF. The notation α^+ used below is for the next larger cardinal after α.

---

1) Over the years, A. A. Fraenkel has written several general expositions on set theory. The most recent is the second edition of his book with Y. Bar-Hilel prepared by A. Levy, *Foundations of set theory*, 1973. This book covers a broad range, not limiting itself to set theory. It contains no proofs, includes many historical notes, and is confined to the relatively elementary part of set theory. To take just one example of the richness in historical details (p. 138); the proposal of considering an impredicative extension of ZF is traced back to *Proc. Nat. Acad. U..S.A.*, vol. 35 (1949), pp. 150—155, followed by J. L. Kelly, *General topology*, 1955, A. P. Morse, *A theory of sets*, 1965, as well as developments by Stegmüller, Tarski and Peterson, etc.

To prove the independence of CH, we take ω_2, say, in M and G be a generic set consisting of conditions from P which consists of all conditions p each of the form $\alpha_1 \notin G, \cdots, \alpha_m \notin G, \beta_1 \in G, \cdots, \beta_n \in G$, where $\{\alpha_1, \cdots, \alpha_m\}$ and $\{\beta_1, \cdots, \beta_n\}$ are finite subsets of ω_2 (in M). It is easy to see that any set C in M consisting of incompatible conditions from P is countable in M. This is sometimes called the countable anti-chain condition.

Lemma. A set C in M of mutually incompatible conditions is countable in M.

Suppose C is uncountable. Since every condition is finite in size, there must be some n such that there are uncountably many conditions in C of size $\leqslant n$. Take the least such n. $n \neq 1$, since there can be only 2 incompatible conditions of size 1, viz. $\alpha \in G$ and $\alpha \notin G$, for some α.

Let C_n be those conditions in C of size $\leqslant n$. Take p in C_n. Since p is finite, there must be some α_1 in P_1 such that there are uncountably many members of C_n which disagree with P_1 at α_1. Call this subset D_1. Let p_2 be an element in D_1. Since the conditions in D_1 all agree on α_1, there must be α_2 such that uncountably many elements in D_1 disagree with p_2 at α_2. Let this subset be D_2. Repeating this process $n + 1$ times, we arrive at some member of C_n with more than n elements.

A cardinal of M is an ordinal α in M for which there is no function f in M which maps a smaller ordinal onto α. When there is such a function in f, α is no longer the smallest ordinal of its cardinality. (It is possible to choose G so that a cardinal number of M is "collapsed" to a smaller cardinal number in $M(G)$.)

The conditions defined for ω_2 in M do preserve cardinals.

Lemma. The cardinals in M remain cardinals in $M(G)$.

Otherwise we have cardinals α and $\beta = \alpha^+$ in M such that there is a function in $M(G)$ mapping β onto α. Take ω_1 and ω_2 in M as α and β. Hence, there is some condition p which forces this. For each $\alpha < \omega_1$, let A_α be the set of β such that there is $q \leqslant p$, $q \Vdash f(\alpha) = \beta$. For each β in A_α, choose a condition p_β, such that $p_\beta \Vdash f(\alpha) = \beta$. $\{p_\beta : \beta \in A_\alpha\}$ is an antichain for each α, because two conditions forcing $f(\alpha)$ to take different values must be incompatible (f being a function). Hence, A_α is countable, and the union

A of A_α over $\alpha < \omega_1$ has cardinality ω_1. $\text{Sup}(A) < \omega_2$. Hence, there must be some β, $\beta \notin A$, contradicting the hypothesis that the mapping f is onto.

Now we can show that the continuum has at least cardinality ω_2 in $M(G)$. Or, $M(G) \Vdash 2^\omega \geqslant \omega_2$.

Let f be any function mapping ω_2 one-one onto the set of limit ordinals $< \omega_2$. For example, $\omega\alpha$ as $f(\alpha)$ would do. Let $T(\alpha)$ be $\{n | f(\alpha) + n \in G\}$. It is sufficient to show $T(\alpha) \neq T(\beta)$ for $\alpha \neq \beta$. Otherwise, suppose $p \Vdash T(\alpha) = T(\beta)$. Let k be a number such that p mentions neither $f(\alpha) + k$ nor $f(\beta) + k$. Let q be obtained from p by adding $f(\alpha) + k \in G$, $f(\beta) + k \notin G$. Then $q \Vdash T(\alpha) \neq T(\beta)$, contradicting $p \Vdash T(\alpha) = T(\beta)$. Hence, $2^\omega \geqslant \omega_2$ in $M(G)$.

8. UNIFICATIONS AND DIVERSIFICATIONS

As mathematical logic develops continually, one is struck by the phenomenon of diversification and it is hard to see the whole structure. We have so far left out various aspects of the technical advances in recent years. To partly compensate for this inevitable partiality, we shall briefly consider a few other aspects and say a few words about the technical literature in the general field. Finally we shall conclude with a vague suggestion of what seems to be the unifying outlook today.

The special aspects we have chosen are three. The early history of proof theory and Hilbert's program makes a good story. Some of the basic concerns of constructivism require a brief (if superficial) mention. The axiom of determinacy turns out to be surprisingly interesting in its power to yield strong results and in its close relation to large cardinal axioms.

8.1 Proof theory and Hilbert's program

Hilbert's study of the foundations of geometry yielded his famous book on the subject in 1899. In the summer of 1900 Hilbert gave his address "Mathematical problems" at the second International Congress of Mathematicians in Paris[1]. In the same year, he proposed an inexact axiom system for classical analysis and asked as the second problem of his famous list how to establish the consistency of axiom systems like this. In his address of 1904, Hilbert for the first time distinguished the situations in geometry and analysis[2]. "In examin-

---

1) See, e.g., C. Reid, *Hilbert*, 1970.

2) For reference to Hilbert's papers see David Hilbert, *Gesammelte Abhandlungen*, vol. 3, 1935 which includes also a review article by Paul Bernays. Some other papers are reprinted in the seventh edition of his *Grundlagen der Geometrie*, 1930. Call these two books A and B. The relevant writings by Hilbert are then: the 1900 list of problems (A), 1900 "Zahlenbegriff" (B), 1904 Heidelberg address (B), 1917 Zürich address (A), 1922 Hamburg and Leipzig addresses (A), 1925 Münster address (B), 1927 Hamburg address (B), 1928 Bologna address (B), 1930 Hamburg address (A), and "Beweis des tertium non datur," *Nachr. Ges. Wiss. Göttingen*, 1931, pp. 120—125.

ing the foundations of geometry it was possible for us to leave aside certain difficulties of an arithmetic nature; but recourse to another fundamental discipline does not seem to be allowed when the foundations of arithmetic are at issue.'' He spoke of a direct proof of consistency and suggests that by formalizing mathematical proofs and trying to prove that all theorems have a common property not shared by nontheorems, the question of proving consistency becomes a combinatorial problem.

It was not until 1917 when Hilbert returned to the subject with a lecture on axiomatic thinking in which the axiomatic method was discussed in a broad setting, the successful axiomatization of logic (considered as including number theory, set theory, and much more) by Frege and Russell was praised, and the consistency question of number theory and set theory was mentioned together with several other rather indefinite ''deep'' questions. Only after this talk, especially since 1920, did Hilbert, with the close collaboration of Bernays, begin to concentrate on proof theory, much stimulated by the opposition which H. Weyl and Brouwer directed against the usual way of proceeding in analysis and set theory.

In the address of 1922, Hilbert began to use the terms proof theory and metamathematics[1]. His finitist or formalist viewpoint was given the dramatic formulation: ''Am Anfang ist das Zeichen.'' The formal system and the intuitive metamathematical reasoning were more sharply distinguished than before. In this and an immediately following address, Hilbert began to outline an approach toward proving the consistency of classical analysis. After these initial explanations, there was the belief that the approach contains the basic ideas needed to get the proof, and these can be brought out with sufficient diligence and concentration[2].

---

1) For example, on p. 169 and p. 174 of the book A, Professor Wang Sianjun points out that Cantor used the term ''metamathematics'' in 1884 (see p. 213 and p. 391 of Cantor's *Abhandlungen*). From the contexts, it appears that the word had been used before and that Cantor was not willing to have it applied to his work.

2) See the observation by Bernays on p. 210 of the book A in speaking of the problem of proving the consistency of number theory and analysis: ''zu ihrer Behandlung hatte man den Hilbertschen Ansatz zur Verfügung, und es schien anfangs, dass es nur einer verständnisvollen und eingehende Bemühung bedürfe, um diesen Ansatz zu einem vollständigen Beweis auszugestalten.''

Attempts to carry out this approach include[1] Ackermann's paper of 1924 and von Neumann's of 1927. It was thought that these proofs are at least adequate for the consistency of number theory, and Hilbert not only asserted this in his addresses of 1927 and 1928, but went on to state that only some forthcoming purely arithmetical elementary lemma would prove the consistency of analysis also.

We have mentioned in Chapter 2 that Gödel got the incompleteness results in 1930 which show that Hilbert's approach cannot work (certainly with the restriction to his finitist position). It may be of historical and heuristic interest to paraphrase here Gödel's own statements on his initial approach.

In the summer of 1930, Gödel began to study the problem of proving the consistency of analysis. He found it mysterious that Hilbert wanted to prove directly the consistency of analysis by the finitist method. He believed generally that one should divide the difficulties so that each part can be overcome more easily. In this particular case, his idea was to prove the consistency of number theory by finitist number theory, and prove the consistency of analysis by number theory, where one can assume the truth of number theory, not only the consistency. The problem he set for himself at that time was the relative consistency of analysis to number theory; this problem is independent of the somewhat indefinite concept of finitist number theory.

To return to Hilbert, it will be of interest to look at the list of open problems which he listed and explained in his Bologna address of 1928. The first problem is the consistency of analysis which he considered essentially solved. He went on to draw the conclusions that the impredicative definitions and Cantor's second class (viz. the countable ordinals) were thereby clarified. The second problem is to extend the consistency proof to functions of real variables and even higher types. He mentioned in particular the consistency question of the axiom of choice (which, as we mentioned in the previous chapter, was later solved more generally in a relative form by Gödel in 1938).

Hilbert's third problem was the question of completeness of form-

---

1)   W. Ackermann, *Math. Ann.*, vol. 93 (1924), J. von Neumann, *Math. Z.*, vol. 26 (1927).

al systems of number theory and analysis. This, as we know (see Chapter 2), has been answered negatively by Gödel in a strong form two years later. The fourth and last problem was the completeness of first order logic, which was answered positively within a year by Gödel. It is indeed remarkable that such a list of basic problems was all settled definitely so soon after the problems were presented publicly, and all by the same person.

From Gödel's incompleteness theorems, it is clear that Hilbert's original finitist methods[1] are not enough to prove even the consistency of the first order Peano arithmetic PA. On the other hand, there is a very direct translation of PA into the formal system of intuitionistic arithmetic discovered independently[2] by Gödel and Bernays-Gentzen. Hence, from the intuitionistic viewpoint, the consistency of PA is not in question. However, it is now generally agreed that even in number theory, intuitionist methods contain finitist methods as a proper part.

In 1935, Gentzen arrived at the first consistency proof of PA

---

1) The most likely explanation of this vague concept is that they correspond more or less to slight extensions of primitive recursive arithmetic (without quantifiers of course). It is a little amusing that while Hilbert often attacked Kronecker (for example, as Verbotsdiktatur on p. 159 of vol. 3 of his *Abhandlungen*), he came closer and closer to Kronecker's position and stated in his address of December, 1930: "Etwa zu gleicher Zeit, also schon vor mehr als einem Menschenalter hat Kronecker eine Auffassung klar ausgesprochen und durch zahlreiche Beispiele erläutert, die heute im wesentlichen mit unserer finiten Einstellung zusammenfällt." (*Math. Ann.*, vol. 104, no. 4). In this same lecture Hilbert introduced an infinite rule which permits one to introduce $\forall xFx$ as a new axiom if for each numeral $\bar{n}$, $F\bar{n}$ is true by elementary calculations. This rule would, for example, "prove" and decide the undecidable sentences of Gödel, but it is no longer a part of any formal system. It is tempting to speculate on Hilbert's motive behind the introduction of this rule. It was three months earlier (September, 1930) when both Hilbert and Gödel were giving lectures at the Königsberg congress. Hilbert gave an address *Naturerkennen und Logik* which concluded with "Wis müssen wissen, wir werden wissen" and Gödel announced his first incompletlness theorem.

2) Gödel presented his result on June 28, 1932 which was only published in 1933. The Gentzen manuscript (with contributions by Bernays) was recieved for publication on March 15, 1933 and later withdrawn when in galley proofs, see *Collected papers of Gerhard Gentzen*, 1969, pp. 51—67, p. 313 (note 13), and p. 315 (note 46).

by using a natural extension of the finitist methods. This was replaced by a different proof in 1936 because objections were raised against the original proof on the ground of an implicit use of the fan theorem. In the next few years other proofs were given by Gentzen, Kalmar, and Ackermann[1]. Since about 1950, these proofs are extended to ramified type theory, and then to predicative or Δ^1_1 analysis and to Π^1_1 analysis. We shall not enter into these results but refer the reader to Schütte's recent book[2]. A somewhat different direction is to prove cut-eliminations in the simple theory of types[3].

A different approach was initiated by Gödel[4] in 1942 and first published in 1958. This uses primitive recursive functionals to interpret logical constants and sentences in intuitionistic arithmetic and then by way of the translation of PA into it, an interpretation of sentences in PA is also obtained. This approach has led to a lot of work on the interrelationship between classical mathematics and intuitionistic mathematics[5].

Of course much of the material considered in Chapter 2 above forms a central part of proof theory as well as of all of mathematical logic.

---

1)   For Gentzen's 1935 and 1936 proofs, as well as his new proof in 1938, see his *papers* just cited. L. Kalmar presented a proof in a letter to Bernays dated Sept 24, 1938 which appeared for the first time in Hilbert and Bernays, *Grundlagen der Mathematik*, second edition, vol. II, 1970, pp. 513—535. Ackermann's proof first appeared in *Math. Ann.*, vol. 117 (1940), pp. 162—194; expositions of this proof are given in the Hilbert-Bernays book just quoted, pp. 535—555 and in *A survey of mathematical logic*, 1962, pp. 362—372. K. Schütte gives another proof in *Math. Ann.*, vol. 122 (1951), pp. 369—389.

2)   Kurt Schütte, *Proof theory*, 1977.

3)   See G. Takeuti, *Proof theory*, 1975.

4)   In *Dialectica*, vol. 12 (1958), pp. 280—287 (the particular double issue is a *Festshrift* for Bernays on his 70th birthday). Gödel lectured on the results in 1942 at Princeton and Yale Universities. E. Artin was present at the Yale lecture. An English translation of the 1958 paper with additions was prepared by Gödel before 1970 but has not yet been published.

5)   See Chapter D5 in *Handbook of math. logic*. In fact, all 8 chapters in part D are under the heading: proof theory and constructive mathematics; Chapter D3 deals with the question of "direct proofs." For more on the different directions in proof theory, see the articles by J.-Y. Girard, P. Martin-Löf, and D. Prawitz in *Proc. 2nd Scand. logic colloquium*, ed. J. E. Fenstad, 1971.

8.2   Constructivism

We mention just a few elementary facts about constructivism and constructive proofs. The study of constructive proofs gives rise to several competing, and mutually incompatible, views of constructive mathematics, intuitionism being the best known among them. Some basic requirements seem to be shared by these different views.

When a conclusion $A \lor B$ is proved, a constructivist demands that the proof must contain either a proof of A or a proof of B and that it must tell him which. Similarly a proof of $\exists xAx$ is constructive only if we can find a particular x, a term in the language, satisfying A. In particular, a proof of $\forall x \exists yAxy$ is constructive only if it gives a method of finding y for each x, for example, a recursive function $f(x)$ such that $\forall xA(x, f(x))$ is provable. The standard contrast is between truth in classical mathematics and proof or provability in constructivism. The law of excluded middle $A \lor \neg A$ is a simple universal law classically but becomes a highly strong principle constructively: for example, as a statement of a faith that all mathematical problems are solvable. In fact, the principle appears to be doubtful if we take Axy as arithmetic statements of the form $\forall zRxyz$, where R is recursive. It is possible to choose R so that $\forall x \exists yAxy$ is classically true but there is no recursive function f so that $\forall xR(x, f(x))$ is true. Hence, for this A, $\forall x \exists yAxy \lor \neg \forall x \exists yAxy$ is unprovable according to some form of constructivism.

An example of constructively unacceptable proof is the following
(1)   There exist two irrational numbers a and b such that a^b is rational.

Proof. Take $(\sqrt{2})^{\sqrt{2}}$. It is either rational or irrational. If it is rational, take $a = b = \sqrt{2}$. If it is irrational, take it as a and $\sqrt{2} = b$.

A simple illustration of the contrast between classical and constructive logics is to consider what is called positive implicational calculus with two axioms and one rule of inference.

A1. $A \supset (B \supset A)$
A2. $(A \supset (B \supset C)) \supset ((A \supset B) \supset (A \supset C))$.
Rule E.   If A and $A \supset B$, then B.

The rule gives us the way of eliminating $\supset$. The two axioms are equivalent to the rule of introducing $\supset$:

Rule *I*.   Whenever we can deduce B from A, we can write down $A \supset B$. From discussions in the literature[1], it is clear that the above system codifies correctly the intuitive meaning of implication as determined by inference.   On the other hand, it is well known that the system does not characterize implication as a classical truth function in the sense that $A \supset B$ is true iff A is false or B is true. For example, $((A \supset B) \supset A) \supset A$ (known as "Peirce's law") is a tautology but not derivable in the above system which is constructively adequate.

If we include also negation, then we get the intuitionistic propositional logic for implication and negation by adding:

A3. $(A \supset \neg A) \supset \neg A$.

A4. $\neg A \supset (A \supset B)$.

To get classical logic, we just add $\neg \neg A \supset A$ which functions like the law of excluded middle. By the way, if we weaken A3 and A4 to a single axiom $(A \supset \neg B) \supset (B \supset \neg A)$, we get what is known as the minimal calculus.

8.3   The axiom of determinacy

We have stated the axiom in Chapter 4 (at the end of 4.2) and mentioned its finite case, which is easily provable, toward the end of 1 in Appendix A.   Apparently this was first proposed as an "axiom" by Hugo Steinhaus between 1959 and 1962 as a simple generalization of the finite case (a sort of infinite De Morgan's law)[2], which, using s as the sequence of values of $\{x_1, y_1, x_2, y_2, \ldots\}$, takes the form:

(1) $\forall x_1 \exists y_1 \forall x_2 \exists y_2 \cdots s \in A \; \bigvee \; \exists x_1 \forall y_1 \exists x_2 \forall y_2 \cdots \neg s \in A$.

---

1)   See, for example, H. B. Curry, *A theory of formal deducibility*, 1957 and Supplement III of the Hilbert-Bernays book cited under footnote 1) on page 160.   For more on constructivism, see D5 in *Handbook of math. logic* and the list of references given there. L. E. J. Brouwer's collected papers on foundations of mathematics appear in 1975.

2)   The first publication is apparently J. Mycielski and H. Steinhaus, *Bull. Pol. Acad.*, vol. 10 (1962), pp. 1—3.

Actually as early as 1953, the general statement was shown to be refutable with the help of the axiom of choice and the simple case of (1) was proved when A is an open set[1].

The major attention has been devoted to projective determinacy (PD). Let us recapitulate the relevant definitions. We have mentioned the fact that the projective hierarchy (of sets of reals) is quite analogous to the analytical hierarchy (of sets of natural numbers). Let us change the notation in Greek letters slightly to use E_1 for the analytic sets in the classical sense, A_n for the complements of E_n sets, E_{n+1} for the projections of A_n sets, and $D_1 = A_1 \cap E_1$ for the Borel sets[2].

8.3.1   Let A be a set of reals (or rather of sets or functions of natural numbers); it is common to take A as a subset of all functions $f : \omega \to \omega$. A 2-person infinite game $G(A)$ of perfect information is associated with A. Player I chooses $n_0 \epsilon \omega$; then Player II chooses $n_1 \epsilon \omega$; then I chooses $n_2 \epsilon \omega$; and so on. Let $f(i) = n_i$. I wins iff $f \epsilon A$. I (or II) has a winning strategy if there is a function g from finite sequences of natural numbers to ω so that I (or II) wins by following the strategy however the opponent plays. $G(A)$ is determined if one of the players has a winning strategy. The axiom of projective determinacy (PD) states that for every projective set A, $G(A)$ is determined. Similarly we can speak of A_n determinacy, D_1 determinacy, etc.

A basic theorem on regularity properties of sets of reals is that determinacy implies much regularity:

8.3.2   If all projective (or all A_n) games are determined, then every projective set (or every set in A_n) is Lebesgue measurable, has the Baire property, and has the perfect subset property.

In some respects, PD serves the function of large cardinal axioms. For many years, Solovay has proposed a program of relating large cardinal axioms to PD. Some of the earlier results related to the com-

---

1)   D. Gale and F. M. Stewart, *Annals of math. studies*, vol. 28 (1953), pp. 245—266.

2)   Compare J. E. Fenstad in *Proc. 2nd Scand. logic colloquium*, 1971, pp. 41—61 and Chapter C8 of *Handbook of math. logic* for more details and reference. D. A. Martin is preparing a monograph on the axiom of determinacy.

parison are[1]:

8.3.3   If there exist measurable cardinals, then all A_2 sets have the regularity properties listed in 8.3.2.

8.3.4   The axiom of determinacy implies that ω_1 is a measurable cardinal.

8.3.5   If measurable cardinals exist, then all A_1 games are determined (briefly, MC implies A_1 determinacy).

Recently Martin has derived D_2 determinacy from the existence of large cardinals. Apparently he has also extended his result to the whole PD by using very large cardinals.

One interesting problem was open for a long time, viz. whether Borel determinacy can be proved in ZF. A surprising result by Friedman made it clear that a different kind of proof is called for[2]:

8.3.6   Borel determinacy cannot be proved in the original Zermelo set theory (i.e., when the definite properties are made explicit in the familiar manner, ZF without the axiom of replacement but with the axiom of separation). In fact, this has been strengthened to the extent that rank ω_1 is necessary.

The proofs of other regularity properties can be carried out in the formal theory of ω and its power set. But 8.3.6 shows that even if we use all the finite types, we cannot prove Borel determinacy. Friedman's result is of interest also in the search for mathematical propositions which are independent of commonly accepted formal theories. It belongs with the independence results of Ramsey type theorems from formal number theory (as reported in the last section of Chapter 2).

In 1975, Martin proved the theorem[3]:

8.3.7   All Borel games are determined. He states that the improved form of 8.3.6 is best possible in the sense that the proof of 8.3.7 requires just the strength that is necessary according to 8.3.6. He promises to give details on this matter in his forthcoming monograph *Borel and projective games*.

---

1)   8.3.3 and 8.3.4 are by Solovay, published in *Foundations of mathematics*, 1969, pp. 58—73 and *Notes* distributed at 1967 summer conference on set theory. 8.3.5 is by D. A. Martin, *Fund. math.*, vol. 66 (1970), pp. 287—291.

2)   H. Friedman, *Annals of math. logic*, vol. 2 (1971), pp. 326—357.

3)   D. A. Martin, *Annals of math.*, vol. 102 (1975), pp. 363—371.

8.4 Comments on the literature in mathematical logic

We have given frequent references to various books and articles in footnotes. It may be of some use to make some general comments here. An interesting point of reference is the recently published *Handbook of mathematical logic* which contains a helpful collection of articles, written by specialists, on different areas currently relevant to more or less routine research. It has of course the disadvantage of the inevitable lack of unity when a book collects together articles by different people. In the areas it covers it offers a good supplement to the present book both in giving more details and in treating topics left out here.

The book consists of 31 chapters written by 34 authors. The table of contents gives some indication of many of the areas in mathematical logic today.

A. Model theory. A1. An introduction to first order logic. A2. Fundamentals of model theory. A3. Ultraproducts for algebraists. A4. Model completeness. A5. Homogeneous sets. A6. Infinitesimal analysis of curves and surfaces. A7. Admissible sets and infinitary logic. A8. Doctrines in categorical logic.

B. Set theory. B1. The axioms of set theory. B2. About the axiom of choice. B3. Combinatorics. B4. Forcing. B5. Constructibility. B6. Martin's axiom. B7. Consistency results in topology.

C. Recursion theory. C1. Elements of recursion theory. C2. Unsolvable problems. C3. Decidable theories. C4. Degrees of unsolvability: a survey of results. C5. α-recursion theory. C6. Recursion in higher types. C7. An introduction to inductive definitions. C8. Descriptive set theory: projective sets.

D. Proof theory and constructive mathematics. D1. The incompleteness theorems. D2. Proof theory: some applications of cut elimination. D3. Herbrand's theorem and Gentzen's notion of a direct proof. D4. Theories of finite type related to mathematical practice. D5. Aspects of constructive mathematics. D6. The logic of topoi. D7. The type-free lambda calculus. D8. A mathematical incompleteness in Peano arithmetic.

Undoubtedly this list of topics gives some indication of what a segment of the working logicians take to be the current scope of mathematical logic. The list is not sufficiently structured to indicate the interrelations or the relative importance of the different topics. It gives no historical perspective. It tends to stress more what the average working logician is concerned with at the present time and tends to pay less attention to more traditional problems. In attempting to be comprehensive, there is a tendency to include much that is of doubtful interest except to the small group of people who happen to do research on those questions. There is something inherent in the process of specialization which creates a pull on the average specialist toward the purer and more autonomous region. A consequence of this bias is the neglect of areas of interest to other disciplines such as the study of computers and computability.

In many subjects oral communication, letters, and preprints have grown in importance as a way of testing ideas and finding out new developments. The more traditional means of journals and proceedings remain of value.

There are at least four journals devoted exclusively to mathematical logic: *Journal of symbolic logic* (1936—), *Archiv für mathematische Logik und Grundlagenforschung* (1955—), *Zeitschrift für mathematische Logik und Grundlagen der Mathematik* (1955—), and *Annals of mathematical logic* (1970—). The last one generally publishes longer papers. The first one occasionally publishes survey papers. Also, the *American mathematical monthly* publishes, from time to time, expository articles in mathematical logic.

During the last ten years or so, there have appeared a number of proceedings of logic colloquia, held in England, Norway, Germany, and other places. These volumes often contain lecture notes summarizing recent developments in special areas.

8.5   Hierarchies and unifications

If one is asked to name some of the most striking open problems in mathematical logic today, the most likely candidates are probably the continuum problem and the tautology problem (often stated in the form whether $P = NP$). Both problems appear to be very far from definitive solutions. There are a number of less determinate central questions or directions: a more concretely applicable model theory, a better understanding of larger cardinals, a closer integration of recursion theory and set theory, a theory of practicable computations, more absolute concepts of definability and provability, etc.

Professor Shen You-ding suggests that one major task is to develop a more comprehensive framework which contains both classical mathematics and constructive mathematics as parts. He is also interested in developing some central system of logic which would localize contradictions, seeing that in the logic commonly in use now any contradiction would make every sentence provable and destroy the distinction between true and false.

Professor Wang Sian-jun asks the question that in the earlier days of mathematical logic Hilbert's program served as a central motivating force for research, what is the central program of mathematical logic today? It seems that one possible answer is the clarification of the structure of all sets (including functions and numbers) with proper measures of complexity.

From this point of view, we may first think of simple functions over the natural numbers. Here a natural question to ask is what functions are practicably computable. Once we make the jump to theoretical computability we arrive on the one hand at the more systematic study in recursion theory and on the other hand at the interconnection with other branches of mathematics through the discovery of decidable theories and unsolvable problems.

At the next stage we come to the area of definability, first extending recursive sets to arithmetical sets. At this stage the further extensions become interlinked. If we use only what is available so far, even with the help of the concept of ordinal numbers, we seem to be able to obtain just the recursive ordinals. Using the notion of first order definability, we can then only get the segment of constructible sets up to recursive ω_1 which already calls for a view from outside of the limited domain of first order arithmetic, perhaps by the use of an inductive definition presupposing more sets.

The simpler and more familiar way is to introduce the concept of power set (the set of all subsets of a given set) and the associated approximation to it by the use of impredicative definitions. Once we add this concept which of course remains in need of a great deal of work directed to its clarification, we arrive at the concept of rank model which permits us to iterate the operation of forming power sets to any transfinite rank α, provided we have the ordinal α available. And we do have many ordinals and cardinals available. For example, just looking at the power set of ω, we already have 2^ω. Since, however, we still do not know how large 2^ω is, we may prefer to use ω_1, ω_2, etc. as indices which are certainly available since the power set of ω has at least ω_1 members, etc. But in this case we seem to get stopped at the rank which is the limit of ω, ω_ω, ω_{ω_ω}, etc.

The familiar way to get beyond is to use the axiom of replacement. But then we get stopped below the first inaccessible number which invites further extensions of the rank hierarchy. At this stage, we begin to enter the range of large cardinals.

The constructible sets provide us with a more orderly interpretation of the rank model in terms of ordinals and first order definability. For a long time, there was no proof of the belief that there exist nonconstructible sets. In 1961 it was proved that if there is a measurable cardinal, then there are nonconstructible sets. Since then it has been found that there are many smaller cardinals which would do this. The general opinion is not to conclude that such cardinals do not exist but rather that there indeed are non-constructible sets. Since, however, the hierarchy L of constructible sets has so many attractive properties, there are serious attempts[1] to find "L-like" models for set theory when large cardinals are present so that L itself can no longer be a model of set theory. The less mathematical efforts to justify and unify the large cardinals are usually made by appealing to reflection principles and elementary embeddings[2].

As we have mentioned before, model theory is primarily concerned with the interaction of sets with first order logic. It does not study sets as such but only use them as interpretations of theories, thereby creating an avenue of interaction with other branches of

---

1) For example, a manuscript *The core model* by Tony Dodd and R. B. Jensen has been in circulation for the last year or two. William Mitchell is working on extensions of these results to get "L-like" models for larger cardinals, the current goal being supercompact cardinals.

2) See R. M. Solovay, W. N. Reinhardt, and A. Kanamori's paper in *Annals of mathematical logic*, vol. 13 (1978), pp. 73—116; and the less technical paper on "large sets", *Logic, foundations of math. and computability*, et. Butts and Hintikka 1977, pp. 309—333. For a general discussion of the concept of set, see *From math. to philosophy*, Chapter 6.

mathematics (especially algebra at its present stage of development). Proof theory by definition deals with proofs which are of course a central traditional concern of logic. From the viewpoint sketched above, model theory and proof theory are concerned with refinements and detailed studies of the structure of sets, functions, and numbers, having to do more with theories and inferences rather than with the existence and properties of sets and functions.

For many years, mathematical logic has been associated with foundations of mathematics. As it acquires more mathematical content, the association becomes less close. It is clear that Frege was interested in giving an exact logical foundation of mathematics. But he felt that his program failed. While *Principia mathematica* created a good deal of interest in mathematical logic, it certainly does not give a logical foundation in the originally intended sense. Hilbert aimed at a different way of securing an evident justification of mathematics by using consistency proofs. Gödel demonstrated the impossibility of such a program. Ever since the main trend is no longer reductionist in spirit but rather shows a willingness to study closely different kinds of infinity (including new ones as they emerge).

Apart from this basic distinctive concern with sets and infinities as such, what remains distinct in mathematical logic (in contrast with other branches of mathematics) seems to be the interest in the language of mathematics (both the attention to theories in model theory and the concern with syntax in proof theory, as well as the study of formal languages such as computer languages), in constructivism, and in new axioms (such as the axiom of determinacy and axioms of new cardinals).

APPENDIX A. DOMINOES AND THE
INFINITY LEMMA

1   Some games of skill

The simplest example is the trivial and futile game of tick-tack-toe: futile because if both players have a moderate amount of experience and intelligence, the game always ends in a draw; trivial because the correct strategies are easy to memorize. The most advantageous move of the first player is to take the center square and then the second player must take a corner square. The rest of the game is more or less automatic. The following typical sequences of moves should be sufficient to illustrate the triviality of the game.

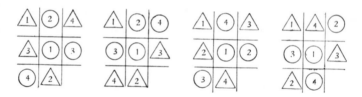

If the second player takes a side position, then the first player always wins by taking any side position which demands an immediate response:

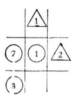

Although the examples convince us that the game is futile, a watertight proof of the fact would be somewhat more tedious. In more serious mathematics the gap between a convincing sketch and a formalized proof can be far greater.

It is rather obvious that if we use a bigger board, say 4 by 4, then the first player has a winning strategy: i.e., a method by which he can always win. Take one of the four center squares and he has three ways of taking another of the four center squares at the next move. The second player can only use his first move to block one of the three ways, and he has no effective response for his second move. For example:

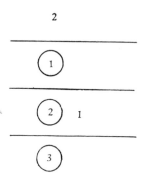

A more interesting generalization is to study any game of skill (not of chance) with two players and with at most three outcomes: draw, the first player wins, the second player wins. The concern is not so much a particular play of a game as the existence or nonexistence of winning strategies. Such considerations can bring out the abstract structure of all games of this type, as will be seen with the help of a slightly less familiar game, that of "nim".

Arrange any number of objects, say six tin soldiers, in three piles, for example: $\alpha 1$, $\beta 2$, $\gamma 3$.

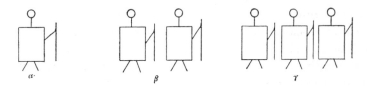

Two players, A and B, draw in turn. Each time each player takes any number of soldiers from any pile. He may take all (remaining) soldiers in a pile or leave some, but he must take at least one soldier each time. He may take from only one pile during any one turn, though he may vary the pile with each turn if he wishes. Whoever takes the last soldier is the winner. Since the (finitely many) soldiers will eventually be exhausted, it is obvious that the game permits no draw. It is of interest that exactly one of the two players has a winning strategy. It is A or B depending only on the sizes of the initial three piles.

In the particular game given above, the second player B, i.e., the player who does not make the first move, has a "winning strategy." Before explaining more exactly this concept, let us give a particular winning play for B. The initial piles are represented by 123, and each move is represented by the number of soldiers taken, preceded by the pile $(a, \beta, \text{or} \gamma)$ from which they are taken.

	Move	Result
Player A	$\alpha 1$	023
Player B	$\gamma 1$	022
A	$\beta 1$	012
B	$\gamma 1$	011
A	$\beta 1$	001
B	$\gamma 1$	000

In this play, B wins because B takes the last soldier, producing the situation 000 under which A can make no more moves. Actually, it is clear that B is going to win when B produces 011, because A can only produce either 001 or 010. More, once A produces 012, B is sure to win provided only he plays correctly. In fact, as soon as B produces 022 (and therefore as soon as A produces 023), there is no way for A to win without some unwise move on the part of B. This is so, because for every move by A after 022, B has a countermove to produce either 011 or directly 000. What is rather more surprising is that A is lost even before the game begins, provided B is sufficiently clever, no matter how much more clever A is. This is the intuitive sense of saying that B has a winning strategy. The following tree is helpful for an explanation of this.

Thus, at the starting situation with 123, player A has six possible moves, $\alpha 1$ (take one soldier from the first pile) or $\beta 1$ or $\beta 2$ or $\gamma 1$ or $\gamma 2$ or $\gamma 3$, resulting in the six situations on level 1 of the tree.

The strategy of player *B* is to choose one counter move for each alternative. And the whole tree shows in a summary fashion that *B* can end up taking the last soldier no matter how *A* plays. Obvious repetitions are omitted according to the convention that we pursue the left branches first. For example, since we have seen on level 4 at the left end that 011 is a winning situation for *B*, we can stop when *B* has produced 011. Similarly with 022, 110, 101. We imagine these omissions which are indicated by dots to be filled.

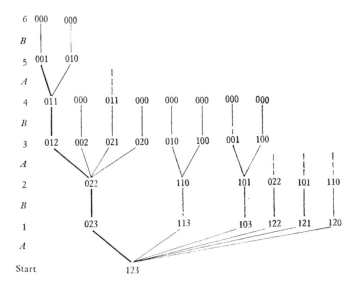

In the above tree, every path from the bottom (the starting situation 123) to the top (the situation 000) represents a play, and, in fact, a winning play for player *B*. For example, the heavy lines give the winning path for *B* that represents the previously described play.

A winning situation for *B* is a situation such that no matter how *A* moves next, *B* can always find a counter move to produce either the final result 000 (which is a special case of a winning situation for *B* because *A* can make no more moves) or another winning situation. More exactly, this is defined by a somewhat complex recursive condition: a situation *S* is winning for *B* if (1) *A* is to make the next move and *S* is 000 or such that for every next move by *A*, there is a move by B which yields a winning situation for B; and (2) B is to make the next move and there is a move by B which yields a winning situation for B according to (1). When A and B are interchanged throughout the definition, we obtain the concept of a winning situation for A.

According to this definition, all the situations in the above are winning for B. Thus, 000 is always produced after a move by B; 001, 010, 002, 020, 100, 001 all are produced after a move by A (so that B is to make the next and winning move). Continuing downwards, we see that 011 is winning for B by levels 4, 5 and 6 in the tree; 022, 110, 101, and finally the starting situation 123, all are winning for B.

Now player X is said to have a winning strategy if the initial situation is a winning one for X. We have seen that in the game beginning with 123, the player B has a winning strategy. From this it follows immediately that if we keep two piles fixed and make the remaining pile bigger in any manner, A would have a winning strategy. For example, given 823, A can reduce it to 123 and become the second player of the old game; similarly A can reduce 173 to 123, 129 to 123, and so on. Hence, we get a result which applies to infinitely many different games.

Can it happen that both A and B have winning strategies? Not surprisingly, the answer is "no". If we are sufficiently patient, we can fatten the tree above by including all legally possible moves at each stage and consider the (finite) totality of all possible games (i.e., paths from 123 to 000). In this fattened tree, there will be winning paths for A and winning paths for B. Since we know there is a winning strategy for B, we wish to prove there cannot be a winning strategy for A. Suppose otherwise. Then we would be able to obtain a subtree in which every situation is winning for A. This is impossible because no matter what first move A chooses to make (on level 1 in the thin tree above, since all possible first moves of A are included there), B has a counter move which anticipates successfully all further moves by A. In fact, the thin tree contains a counterexample to every winning tree for A must at every stage permit B to make any next move he wishes. If the reader dislikes this sketchy abstract argument, he might enjoy attempting to construct a winning tree for A.

Unlike many other games, the game of nim never produces a draw. It follows that for each given set of three piles, either Player A or Player B must have a winning strategy. Thus, if A has no winning strategy, then for every first move of A, there is a move by B for which A cannot win if B continues to play properly. This, however, means that B has a nonlosing strategy. Since there is no draw, this is also a winning strategy for B. In fact, we can always calculate whether A or B has a winning strategy.

Suppose we are given three piles with 234, 10^7, 2729 tin soldiers. It is theoretically possible to tabulate all the possible games with these three piles and then by inspection tell whether A or B has a winning strategy. But no one is willing or able to do such a calculation mechanically. Once such a thankless task is entered upon, shortcuts will inevitably suggest themselves. A systematic search for shortcuts which are applicable in a general manner would be doing mathematics in a most respectable sense: to make calculations easier or, more impressively, to achieve economy of thought. In the game of nim, we happen to have a simple recipe by which we can decide which player, if clever enough, can always win.

Suppose a, b, c are the numbers of the tin soldiers in the three piles and A is to play first. We represent a, b, c as polynomials in terms of 2 (in other words, in the binary notation) and sum up their coefficients:

$$
\begin{aligned}
a = 27 = 2^4 + 2^3 + 2 + 1 &= 11011 \\
b = 37 = 2^5 + 2^2 + 1 &= 100101 \\
c = 46 = 2^5 + 2^3 + 2^2 + 2 &= \underline{101110} \\
& \ 212222
\end{aligned}
$$

The recipe is simply that A can always win if and only if either 1 or 3 occurs in the sum of coefficients. If 1 or 3 occurs, A can so play the game that after each of his moves, 1 or 3 no longer occurs in the sum of coefficients. In the example given, A need only reduce the first pile from 27 to 11, taking away 2^4. In the general case, if 1 does not occur in the sum, it is easy to reduce all the 3's to 2's by working on any one pile. Otherwise A can work on the pile which contributes a highest 1 (i.e., the coefficient of a highest power of 2) to an odd digit in the sum of coefficients, taking care to even out the 1's contributed by one of the other two piles. In particular, a 1 contributed by another pile alone is brought up to a 2. After A's move , B will inevitably destroy the balance so that some 1 or 3 will come into the sum of coefficients. Hence, if A always makes the right move, he can maintain a sum of coefficients with 0's and 2's only, all through the game. The interested reader can undoubtedly work out the details for himself.

The game of nim serves to illustrate two points. (1) Mathematical considerations can reduce a lengthy calculation procedure to a more elegant and practically more feasible one. (2) Simple special cases can help the study and understanding of more general situations. In particular, the considerations above apply to a wide class of games with two players: if there is a finite bound to the total number of possible sequences of moves and there is no "draw", then exactly one of A and B has a winning strategy so that one might say that such games are unfair. If draw is permitted, then the only other possibility is that both have a nonlosing strategy, and one might say that such a game is futile.

If we represent each game by a tree so that each node represents a position which is directly connected to all the positions resulting from the next move, then our main result so far may be stated thus: Every game is either futile or unfair if its tree contains no infinite paths (i.e., there are no endless games) and only finitely many branches come directly from each node (i.e., there are finitely many legal moves at each stage). This is so because the two conditions guarantee that the total number of possible complete plays is finite[1].

The theorem does not apply directly to the game of chess because there are no precise rules to prevent endless games. It would, however, hold even in the stronger form that chess is always unfair

1) An early mathematical study of gemes of skill is contained in E. Zermelo, "Über eine Anwendung der Mengenlehre auf die Theorie des Schachspiels," *Proceedings of the Fifth Int. Cong. of Mathematicians*, 1912, Vol. 2, pp. 501—504.

if a rule were included according to which the player who is confronted by a position that he has previously encountered in the same play and who makes the same move a second time is defeated. This is so because there are only finitely many possible positions say t altogether, and within $2t + 1$ moves, at least one of the players must have repeated a previous move. The rule is, however, not very reasonable and it has not been adopted.

According to the so-called German rule, a game of chess is a draw if the same sequence of moves occurs three times in succession. It is easy to show that this rule does not exclude endless games by using an infinite sequence of occurrences of two symbols only, constructed first by A. Thue (see next section), in which no (finite) subsequence occurs three times successively.

It is of interest to compare the futile game of tic-tac-toe with the unfair game of nim in order to understand the general situation with finite games of skill. We consider the following general situation. There are two players A and B such that A makes the first move and each player can have at most n moves before the game finishes. At the end of the game, there is a set of possible configurations E which contains a subset C as the winning configurations for A and a subset D as the winning configurations for B. In the case of nim, $E = C \cup D$ or $D = E - C$, while in the case of tic-tac-toe, this is not so because there is another subset F of E consisting of configurations which are not winning ones for either A or B.

Theorem. If $D = E - C$ in a game, then the game is unfair.

There is a very elegant proof for this. To say that A has a winning strategy under the assumption that each player makes at most n moves, we can use the following sentence:

(1) $\exists x_1 \forall y_1 \exists x_2 \forall y_2 \cdots \exists x_n \forall y_n$ (the final configuration belongs to C)

Similarly, the following sentence says that B has a winning strategy:

(2) $\forall x_1 \exists y_1 \forall x_2 \exists y_2 \cdots \forall x_n \exists y_n$ (the final configuration belongs to D).

Since $D = E - C$ and E is the set of all possible end configurations, belonging to D is equivalent to not belonging to C. Moreover, it is familiar from first order logic that, for example, $\neg \exists x \forall y Rxy$ is equivalent to $\forall x \exists y \neg Rxy$. Hence, (2) is just the negation of (1). Therefore, by the law of excluded middle (i.e., the law $p \lor \neg p$), we get: (1) $\lor$ (2). That is to say, either A has a winning strategy or B has a winning strategy. Hence, the game is unfair and the Theorem is proven.

One is tempted to generalize the Theorem to infinite games by using an infinite string of quantifiers. The result is then no longer a consequence of simple logic but in fact contradicts the general axiom of choice. It is known as the axiom of determinacy as stated in chapter 4 (at the end of 4.2).

A natural generalization of tic-tac-toe is to consider games of four-in-a-row, five-in-a-row, etc., and, in general, games of m-in-a-row with a board of size n by n, for arbitrary m and n. Clearly, there is no game if $m > n$. But for what values of m and n is the game futile or unfair in favor of the first player? What happens if the board is infinite (or rather "potentially" infinite, i.e., unbounded)? These are mathematical problems but not, probably, very interesting ones: it is unlikely that very clever proofs of quite general results can be found, or that the answers will have significant applications inside or outside mathematics.

In practice, a fairly intricate game of five-in-a-row is rather widely played with a board of the size between 10 by 10 and 19 by 19. In China, when people find the game of Wei Qi (the surrounding game or go) too demanding, they use the same board and pieces to play five-in-a-row. Apparently, nobody plays four-in-a-row or seven-in-a-row with the 19 by 19 board: obviously the former is unfair and probably the latter is futile.

Xiang Qi (Chinese chess), which also uses 32 pieces, but 10 instead of 16 pawns and a board of size 9 by 10 instead of 8 by 8, seems to be a more attractive game in that one does not feel so crowded. It is surprising that it has attained no popularity in the West.

A simple game is Si Zi Qi (the game with four pieces). It uses a 4 by 4 board. Each of the two players has four pieces initially placed at the opposite edges. A player P destroys a piece of his opponent by a move which creates a situation with exactly three consecutive pieces in a row or a column such that P has two neighboring pieces. The player whose pieces are reduced to one loses the game. It is not obvious whether the game is futile or unfair.

2 Thue sequences[1]

Given an alphabet of n symbols $\{a_1, \ldots, a_n\}$, we ask whether it would be possible to make up an infinite sequence with these symbols in which no finite part immediately repeats itself. For $n = 2$, it is easy to see that this is impossible, in fact:

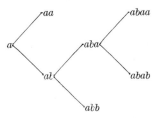

Thus, beginning with a, we get a repetition with any string of length 4 or more; similarly, if we begin with b. Thue proves:

Theorem 1. For $n > 2$, we can find infinite sequences built up from n symbols such that they contain no part of the form UU, where U is a finite (nonempty) string. For $n = 2$, we can find infinite sequences which contain no part of the form UyV, where y is a single symbol, such that $Uy = yV$.

For $n = 2$, let a and b be the two given symbols. The second half of the theorem follows from the more explicit assertion below.

Lemma 1.1 The union of the sequences obtained from a by successively replacing a by ab, b by ba contains no part UyV with $Uy = yV$:

$$R_1(a, b) = a, \quad R_{n+1}(a, b) = R_n(ab, ba), \quad R(a, b) = R_n(a, b).$$

The sequence $R_1, R_2, \ldots$ is clearly:

$$a, \ ab, \ abba, \ abbabaab, \ abbabaabbaababba, \ \cdots$$

By the union $R(a, b)$, we mean the unique sequence R such that for every n, R_n is an initial string of R. It is easy to verify that $R_m(a, b)$ is an initial segment of $R_n(a, b)$ for $m < n$.

In order to prove Lemma 1.1, it is sufficient to prove that for

---

1) For Thue sequences and other related material, see Axel Thue, ''Über die gegenseitige Lage gleicher teile gewisser zeichenreihen,'' *Skrifter utgit av Videmskapsselskapet i Kristiania, 1912, Mat.-Nat. Klasse*, No. 3, 1913, 67 pp.

each n, R_n contains no UyV with $Uy = yV$. Thus, if R contains such a part UyV, the part must occur in a finite initial string and therefore is contained in some R_n. Let P be the property of not having a part UyV with $Uy = yV$. Then, since R_1 certainly has the property P, 1.1 is reduced to:

Lemma 1.2   For every n, if R_n has P, then R_{n+1} also has P.

Suppose now $R_{n+1}(a, b)$, i.e., by definition, $R_n(ab, ba)$ contains a part UyV such that $Uy = yV$. Either $UyV = yyy$, or there is a string K, $UyV = yKyKy$. The first alternative is impossible because no matter what $R_n(a, b)$ is, $R_n(ab, ba)$ cannot contain aaa or bbb as a part. If the second alternative holds, then, depending on whether the first symbol of UyV occupies an even or odd position in R_{n+1}, either $KyKy$ or $yKyK$ must be a string of ab's and ba's which corresponds to a string in $R_n(a, b)$.

If K has an even number of terms, then K and yKy must both be strings of ab's and ba's . Now this is impossible, because by adding two y's to K, in which there is the same number of a's and b's, the balance is destroyed.

Consider now the alternative that K has an odd number of terms. If $KyKy$ is a string of ab's and ba's, then

$$UyV = yKyKy = yNxyNxy,$$

where $x \neq y$ and N is a (possibly empty) string of ab's and ba's. But then UyV must be preceded by x in R_{n+1}, and $xyNxyNxy$ must be a part of $R_{n+1} = R_n(ab, ba)$, obtained by substitution from a part of $R_n(a, b)$, contrary to the assumption that R_n has the property P. Similarly, if $yKyK$ is a string of ab's and ba's, $yxNyxNyx$ must be a part of R_{n+1}. This completes the proof of Lemma 1.2 and Lemma 1.1.

Exercise 1.   If $S_1 = abba$, $S_{n+1}(a, b) = S_n(abba, baab)$, explain the sense in which this would determine a "two-way infinite Thue sequence". Prove your conclusion.

From Lemma 1.1, it follows immediately that:

Corollary 1.3   For $n \geq 2$, there exist infinite sequences such that there are no parts of the form UUU, with U nonempty; in particular, when the symbols are a and b, there are no parts aaa or bbb.

If U is a single symbol y, then yyy satisfies the property P. Otherwise, let $U = yK$, then $UUU = yKyKyK$, and $yKyKy$ satisfies the property P.

This can be applied directly to produce endless chess games under the German rule mentioned above. It is easy to find two moves for each player such that each move can follow either move of the other player. And then each player is free to create the Thue sequence with his two moves labeled a and b, without producing a draw.

Exercise 2.   Find a direct proof of Corollary 1.3 without appeal to Lemma 1.1.

A two-way infinite Thue sequence can be applied to design a pattern of a's and b's on the plane in which there is no "torus". We put the sequence horizontally on the plane and copy each symbol across the diagonal bisecting the first and third quadrants. Then there can be no rectangle block such that left and right edges agree, top and bottom edges agree.

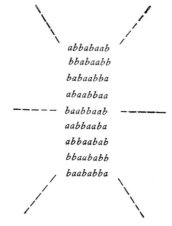

Suppose there were such a block, say,

$$x_4 \; x_3 \; x_2 \; x_1$$
$$x_3 \; x_2 \; x_1 \; y_1$$
$$x_2 \; x_1 \; y_1 \; y_2$$

Then $x_4x_3x_2x_1 = x_2x_1y_1y_2$, and $x_4x_3x_2 = x_2x_1y_1$.   Hence, the part $x_4x_3x_2x_1y_1$ of the Thue sequence would violate the property P.

For the half of Theorem 1 when $n > 2$, it is sufficient to consider three symbols a, b, c.

Lemma 1.4   The sequence T obtained from the sequence R in Lemma 1.1 by substituting ca for aa, cb for bb contains no part of the form DD.

Consider the even terms of R: R_2, R_4, etc. We observe that after the replacement, only four 4-letter sequences occur:

$$A = acba, \quad B = bcab, \quad C_1 = acbc, \quad C_2 = bcac.$$

We distinguish different occurrences of c and use c_2 in a context acb, c_1 in a context bca. Define now T by:

$$T_1(a, b, c_1, c_2) = ac_2babc_1ab$$
$$T_{n+1}(a, b, c_1, c_2) = T_n(A, B, C_1, C_2).$$

E.g., $T_2(a, b, c_1, c_2) = ac_2babc_1ac_2bc_1abac_2babc_1abac_2bc_1ac_2babc_1ab$.

Suppose now DD occurs in T_n and M is obtained from D by substituting A, B, C_1, C_2 for a, b, c_1, c_2, then MM occurs in T_{n+1}, and M begins with a or b. Suppose $M = xN$ ($x = a$ or b). We then have

$$xNxN \text{ in } T_{n+1}.$$

The letter following $xNxN$ must also be x, as otherwise if the last big letter in M is A, B, C_1, or C_2, it would be followed by C_1, C_2, A, or B, but we can only have AC_2B and BC_1A. Hence we have

$$xNxNx \text{ in } T_{n+1}$$

which would yield

$$xUxUx \text{ in } R_{2n+3}$$

The last contradicts the property of the sequence R.

Exercise 3.   Let $P_1(a, b, c) = abc$, $P_{n+1}(a, b, c) = P_n(abcab, acabcb, acbcacb)$. Prove that the union P of these strings contains no part of the form DD.

3   The infinity lemma

Instead of adopting the German rule, we may ask whether it is possible to devise rules which would exclude endless games yet at the same time impose no finite bound on the length of a complete play. Since at each stage there are only finitely many permissible moves, the infinity lemma tells us this is impossible:

Theorem 2. The infinity lemma. If there are infinitely many connected branches in a tree and there are only finitely many branches from each node, then there is an infinite path[1].

Take first the bottom 0 of the tree. Since there are altogether infinitely many branches and only finitely many branches directly from 0, at least one of the nodes 11, 12, ..., $1n_1$ on the next level must be the bottom of a subtree with infinitely many branches.

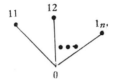

Let it be $1i_1$. By hypothesis, there are only finitely many branches directly from $1i_1$. Hence, on the next level, one of the nodes above $1i_1$ must be the bottom of an infinite subtree. Repeating this argument, we see that on every level there is at least one node which is the bottom of an infinite subtree and these bottoms together determine an infinite path in the original tree.

In order to stress the nontrivial character of the lemma, we may consider a tree in which there are infinitely many branches through the bottom 0 leading to 11, 12, 13, etc., such that $1n$ begins a path of length n.

---

1) The infinity lemma is commonly credited to D. König, "Über eine Schlussweise aus dem Endlichen ins Unendliche." *Acta Litt. Sci.*, Szeged, vol. 3 (1927), pp. 121—130. Somewhat earlier, L. E. J. Brouwer had introduced a more subtle form of this lemma under the name "fan theorem" in "Bewis, dass jede volle Funktion gleichmässig stetig ist," *Proceedings of the Academy of Sciences at Amsterdam*, vol. 27 (1924), pp. 189—193. A generalization of the lemma was introduced by F. P. Ramsey in "On a problem of formal logic," *Proc. London Math. Soc.*, vol. 30 (1929), pp. 338—384.

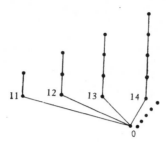

Even though this tree has altogether infinitely many branches and there is no finite bound on the length of the paths, it contains no infinite path.

A direct application of the infinity lemma is that if mankind never perishes, there must exist somebody such that at any future time some descendant of his will be alive.

A less direct and somewhat artificial application is to the coloring of maps. Given any type of surface, if a map with any (finite) number of countries can be colored with p colors so that no two countries with a common boundary line get the same color, then by the infinity lemma, a map with infinitely many countries can also be colored with p colors. Consider only the case when the surface is a plane. It is now known that every map requires only 4 colors and we can take $p = 4$. We wish to argue that the same value p suffices even if we allow infinitely many countries.

Take four colors, say, red, yellow, green, and white. For any (finite) number n of countries, 'there are a finite number of ways of coloring a map of n countries with the four colors. Imagine a map with infinitely many countries $C_1, C_2, C_3, \ldots$ and a tree in which each node on the nth level represents an acceptable coloring of $C_1, \ldots, C_n$ with the four available colors. Suppose further a node of level n is joined to one of $n + 1$ if and only if the colors of $C_1, \ldots, C_n$ are the same in the two coloring. In this way,

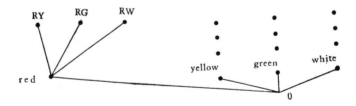

we obtain a tree satisfying the hypothesis of the infinity lemma. Hence, there must be an infinite path which represents a satisfactory coloring of the infinite map.

Of course, there are acceptable colorings of C_1, ..., C_n, C_{n+1} each of which certainly yields a satisfactory coloring for C_1, ..., C_n if we disregard C_{n+1}.

Another application uses a little theory of real variables but is quite understandable without the benefit of any course in that area. Let E be a closed subset of the interval $(0, 1)$, i.e., every x in E satisfies $0 < x < 1$, and every sequence of points in E has a limit in E, and I be a set of intervals such that every point of E lies inside some interval in I. Then there exists a number n such that when $(0, 1)$ is divided into 2^n equal intervals, each such interval which includes a point of E is contained in an interval in I.

Suppose the assertion is false. Then for every n, some of the intervals $\left(0, \dfrac{1}{2^n} \right)$, $\left(\dfrac{1}{2^n}, \dfrac{2}{2^n} \right)$, $\cdots$, $\left(\dfrac{2^n-1}{2^n}, 1 \right)$ contains a point P_n in E but is not included in any interval belonging to I. Since as n is increased by 1, the intervals are halved, we can represent all the intervals by the full binary tree:

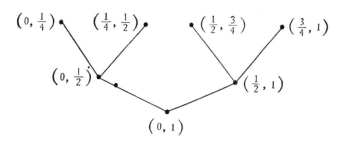

By our assumption, those intervals which contain points of E but are not included in any interval I_j of I form an infinite subtree of the above tree. Hence, by the infinity lemma, there is an infinite path, i.e., there is a sequence of intervals $J_1, J_2, \cdots$ of length 1, $\dfrac{1}{2}$, $\dfrac{1}{2^2}$ such that each interval J_i is not included in any interval of I but contains some point P_i belonging to E. Since E is closed, lim

$P_i = P$ belongs to E. By hypothesis, every point in E, in particular, P is inside some interval I_j in I. But then there exists an integer k such that $\left(P - \dfrac{1}{2^k}, P + \dfrac{1}{2^k} \right)$ is contained in I_j. Hence, there exists an integer m, such that $\left(\dfrac{m}{2^{k+1}}, \dfrac{m+1}{2^{k+1}} \right)$ is J_{k+1} but is contained in I_j, contrary to the construction of the sequence $J_1, J_2, \cdots$ of intervals.

A less intuitive statement of the infinity lemma would run as follows. If there is an infinite sequence $Q_1, Q_2, \ldots$ of disjoint finite sets of ordered pairs of points such that the first point of each pair in Q_{i+1} $(i = 1, 2, \ldots)$ is the same as the second point of some pair in Q_i, then there is an infinite sequence of points $P_1, P_2, \ldots$ such that (P_i, P_{i+1}) belongs to Q_i, for every i.

4   A solitaire with dominoes (tiling problems)

This game was at first formulated in the process of studying the decision problem of first order logic and has turned out to be a convenient tool for relating logical formulas to Turing machines[1]. The game yields a number of abstract problems which have a measure of elegance and can be employed to illustrate, for example, applications of the infinity lemma and Thue sequences.

Assume we are given a finite set of square plates (the dominoes) of the same size (say, all of the unit area) with edges colored, each plate in a different manner. Suppose further that there are infini-

---

1) See *Bell Systems technical journal*, vol. 40 (1961), pp. 1—41. Since the introduction of these problems in 1960, various ramifications have been studied by different people. Some of these studies are considered elsewhere in this book. We mention here just a few more intuitive discussions. (1) *Scientific American*, vol. 213, no. 5 (November, 1965), "Games, logic and computers", pp. 98—106. (2) Ronald L. Graham and Michael R. Garey, "The limit to computation", *Eneyc. Britannica year book*, 1977, pp. 172—185. (3) Martin Gardner, "Extraordinary nonperiodic tiling that enriches the theory of tiles", *Scientific American*, vol. 236, no. 1 (January, 1977), pp. 110—121. (4) Raphael M. Robinson, "Undecidable tiling problems in the hyperbolic plane", *Inventiones Mathematicae*, vol. 44 (1978), pp. 259—264. In (3) and (4) these dominoes or tiles are referred to as Wang tiles.

tely many copies of each domino (domino type). We are not permitted to rotate or reflect a domino so that, e.g., the following three dominoes are regarded as of different types:

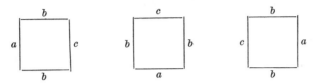

Suppose now we wish to cover up the whole infinite plane with such dominoes so that all corners fall on the lattice points, or, in other words, as with tiles in a bathroom, two dominoes either meet on a corner only, or meet on a whole edge, or do not meet at all. For aesthetic and other reasons one may wish to require that the dominoes of different types be arranged in different patterns. We shall, however, confine ourselves to one very simple requirement, viz., that two adjoining edges must have the same color. Under this restriction, we can speak of solving any given set of dominoes.

4.1 A set of dominoes is said to be solvable (in the infinite plane) if and only if there is some way of covering the whole plane by dominoes from the set.

Thus, in the above example of three dominoes, if a, b, and c are distinct colors, then the set consisting of the first two dominoes has no solution, while that consisting of the first and the third dominoes is solvable because we can join the sides having the color c:

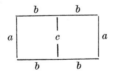

It is easy to see that the same block of two squares can be repeated in every direction since the bottom agrees with the top in color, and the left edge agrees with the right.

Since such blocks are very useful, they are given a special name:

4.2 A torus of a given domino set is a rectangle consisting of copies of some or all dominoes of the set such that: (a) adjoining edges always have the same color; (b) the sequence of colors on the bottom edge agrees with that on the top edge; (c) the left edge agrees with the right edge.

It is obvious that every set with a torus is solvable. Moreover, it has a periodic solution. It is not easy to define in general what a periodic solution is, but it is easy to concede that a solution consisting of only repetitions of a torus is periodic. To fix our concepts, we shall arbitrarily restrict the notion of "periodic" to this particular case:

4.3   A solution of a domino set is periodic if and only if there is a torus T such that by viewing T as a single domino, the solution is made entirely of copies of T.

4.4   Every domino set with a torus is solvable and has indeed a periodic solution.

4.5   If either all rotations or all reflections of a domino is permitted, every domino set is solvable.

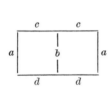 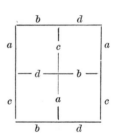

It is easy to get examples which have both periodic and non-periodic solutions.

4.6   *Example* P_1, a set consisting of four domino types:

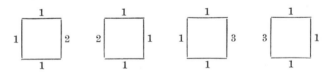

This set has two smallest tori:

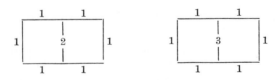

Since these are the same on the outside, we can combine them in any way to get solutions. If we use one torus only, we get of course periodic solutions. If we mix the two types, call them A and B, we then get both periodic and nonperiodic solutions. In fact, since we can distribute A's and B's in any way we like, we get as many solutions as there are binary infinite sequences, i.e., as there are real numbers.

4.7   The set P_1 has as many distinct solutions as there are real numbers.

There are certain sets whose solutions are not all periodic in the sense of 4.3 but are all highly regular.

4.8   *Example P_2:*

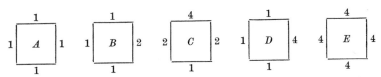

The solutions of P_2 include: (a) all A's (b) all E's. If any of the dominoes B, C, D occurs in a solution, then we get a unique solution as follows:

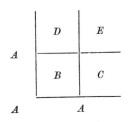

If B does not occur but C occurs, we have:

$$\frac{\frac{E}{C}}{A}$$

If B does not occur but D occurs, we get:

$$A \mid D \mid E \, .$$

4.9 The set P_2 has exactly five solutions as given above.

For certain purposes, it may be desirable to extend the concept of "periodic" so that all solutions of P_2 are periodic in the extended sense. We shall, however, make no such extension but adhere to the narrower notion of 4.3.

4.10 *Example P_3*:

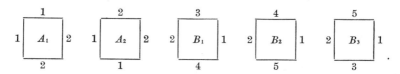

It is easy to see that this set has a smallest torus of size 6 by 2. The example also illustrates how we can get solvable domino sets with large periods.

4.11 *Example P_4*:

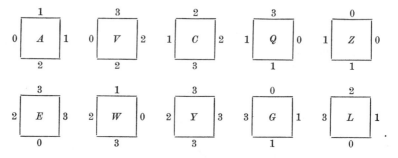

As we noted before in 4.5, permitting rotations or reflections would make the problem of solving domino sets trivial. If, however, one uses regular hexagons instead of squares and allows reflections and rotations, the problem of solving hexagon sets is equivalent to the original problem with squares. Roughly speaking, we can then use certain edges to fix the orientation so that rotations etc. of one hexagon in general forces similar transformation of other hexagons.

4.12 Given a set of square plates, we can find effectively a set of regular hexagons such that there is a one-to-one correspondence between the solutions of the two sets. Conversely, given a set of regular hexagons, we can also find effectively a corresponding set of squares.

Given a set of squares $A_1, \ldots, A_n$, we introduce a set of $2n+1$ hexagons with $6+n$ new colors as follows. A "cementing" hexagon:

Suppose A_i is:

Introduce two hexagons with a new color a_i:

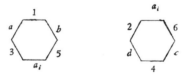

It can be seen that on account of the cementing piece, each pair of hexagons can only be used as a unit thus:

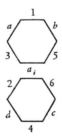

Conversely, given a set of regular hexagons $H_1, \ldots, H_m$, we give one block of three squares for each of the 12 positions for each

H_i, using two new colors for each position. For example, if H_i is

we use for the particular position the obvious combination of the three squares:

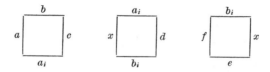

The letter x indicates a new color used everywhere.

Hence, we get a set of $36m$ squares with $24m + 1$ new colors.

Detailed proofs of the two halves of 4.12 are omitted.

In connection with certain decision problems, it is convenient to use a formulation that colors corners instead of edges. For example, if we have the set of "matrix" types:

$$\begin{pmatrix} a & a \\ a & b \end{pmatrix}, \qquad \begin{pmatrix} a & b \\ b & a \end{pmatrix}, \qquad \begin{pmatrix} b & a \\ a & a \end{pmatrix},$$

then we get a solution by the following torus:

$$\begin{pmatrix} a & a \\ a & b \end{pmatrix} \begin{pmatrix} a & b \\ b & a \end{pmatrix} \begin{pmatrix} b & a \\ a & a \end{pmatrix}$$
$$\begin{pmatrix} a & b \\ b & a \end{pmatrix} \begin{pmatrix} b & a \\ a & a \end{pmatrix} \begin{pmatrix} a & a \\ a & b \end{pmatrix} \quad \text{or briefly} \quad \begin{pmatrix} a & a & b & a \\ a & b & a & a \\ b & a & a & b \\ a & a & b & a \end{pmatrix}.$$
$$\begin{pmatrix} b & a \\ a & a \end{pmatrix} \begin{pmatrix} a & a \\ a & b \end{pmatrix} \begin{pmatrix} a & b \\ b & a \end{pmatrix}$$

Exercise 4. Prove that the matrix formulation is equivalent to the domino formulation in the sense that given a set of the one, we can find a set of the other with exactly corresponding solutions.

In 2, we have applied Thue sequences to obtain a pattern of a's and b's on the plane, that contains no "torus." It is possible to modify the construction to get a set of dominoes which has a solution that contains no torus.

In order to find a domino set which has such a solution, we have to represent a and b by several dominoes, taking into consideration the two neighboring squares in the same row. Since we do not permit aaa or bbb, we need only these pieces:

$$a\;\begin{array}{c} a \\ \boxed{\begin{array}{c} a \\ a\ b \end{array}} \\ a \end{array}\;b \qquad a\;\begin{array}{c} b \\ \boxed{\begin{array}{c} a \\ b\ a \end{array}} \\ a \end{array}\;a \qquad a\;\begin{array}{c} b \\ \boxed{\begin{array}{c} a \\ b\ b \end{array}} \\ a \end{array}\;b \qquad b\;\begin{array}{c} a \\ \boxed{\begin{array}{c} b \\ a\ a \end{array}} \\ b \end{array}\;a \qquad b\;\begin{array}{c} a \\ \boxed{\begin{array}{c} b \\ a\ b \end{array}} \\ b \end{array}\;b \qquad b\;\begin{array}{c} b \\ \boxed{\begin{array}{c} b \\ b\ a \end{array}} \\ b \end{array}\;a$$

The required solution is obtained if we replace a and b in the previous solution by domino types $_i a_j$ and $_m b_n$ according to the left and right neighbors (i and j, or m and n).

Exercise 5. Show that any torus in the solution would yield enough identifications in the previous covering of the plane by a's and b's to violate the property of Thue sequences.

Of course, this set has also periodic solutions, e.g., that from the torus[1]

$\begin{array}{c} a \\ b\ b \end{array}$	$\begin{array}{c} b \\ a\ a \end{array}$
$\begin{array}{c} b \\ a\ a \end{array}$	$\begin{array}{c} a \\ b\ b \end{array}$

5   The infinity lemma applied to dominoes

It is quite natural to apply the infinity lemma to prove proper-

---

1) In the original paper of 1961, the problem was proposed to find out whether there exists a set of tiles which has solutions but no periodic solutions. The question was answered in the positive in 1964 by Robert Berger who constructed a solvable set without torus of 104 tiles. Ever since, there has been a sort of competition to obtain smaller such sets. In 1966, H. Lauchli sent me a set with 40 tiles. Several years later R. M. Robinson was able to find a set of 35 tiles. In 1975, Roger Penrose found a small solvable set of polygons without torus; using these generalized tiles, Robert Ammann found in 1977 a set of 16 square tiles which force nonperiodicity. In a recent letter, Professor Robinson reported: "I checked the set of Wang tiles by Robert Ammann. After that, Ronald Graham told me that [John Horton] Conway had reduced the number to 12, but I have not seen Conway's tiles."

ties about domino sets. This is illustrated by a very direct application.

5.1   For a given domino set P, if, for every n, there is a solution of size n by n, then P has a solution.

We consider partial solutions of size $2n - 1$ by $2n - 1$ for $n = 1$, 2, ... and make a tree such that a block K of size $2n - 1$ by $2n - 1$ leads directly only to blocks of size $2n + 1$ by $2n + 1$ with K in the center. The hypothesis and the fact that any part of a partial solution is also a partial solution ensures an infinite tree. Hence, the infinity lemma yields an infinite path in the tree which represents a solution of P.

5.2   A domino set is solvable over the whole plane if and only if it is solvable over a quadrant.

If it is solvable over the plane, we can of course get a solution over a quadrant by deleting the other quadrants from a given solution. Conversely, if it is solvable over a quadrant, then it has a solution of size n by n for every n. Hence, by 5.1, it has a solution over the whole plane.

5.3   Given a domino set P for which it is possible to form two adjacent infinite rows such that for every m, any 1 by m block occurring in the top row also occurs in the bottom row. Then the set P is solvable.

Let A be the given top row, B be the given bottom row. For each m, and a 1 by m block, C_m in A, there is a D_m such that D_m can be correctly put on top of C_m because C_m also occurs in B. Repeating this process with D_m, and so on, we can obtain a partial solution of size m by m. Hence, by 5.1, the set P has a solution.

5.4   A solvable set is said to be minimal solvable if in every solution every domino of the set occurs. A solution is minimal relative to a set M of blocks if there is no solution in which only a proper subset of M occurs.

5.5   Given a minimal solvable set, there exists an integer n such that every domino of the set occurs in every n by n block in every solution.

Assume 5.5 false and we have for every n some solution S and some domino D_i such that there is an n by n block not containing D_i. If the number of dominoes is k, make k trees $K_1, \ldots, K_k$ as in

5.1 so that K_i includes all the $2n-1$ by $2n-1$ $(n=1,2,\ldots)$ blocks which does not include D_i in some solution. By our assumption, at least one of the trees, say K_i, must be infinite. This determines a solution in which D_i does not occur at all, contrary to the hypothesis that the given set is minimal solvable.

5.6 Given a solution S of a domino set P and a finite set K of finite blocks occurring in S, there is a subset M of K such that there is a solution T of P which is minimal relative to M.

Consider every subset of K which contains all blocks in K occurring in some one solution of P. Take the set of all such subsets: this set is not empty since K belongs to it, and it is finite because K is finite. Hence, it has minimal members. Each minimal member M determines a solution T of P which is minimal relative to M.

5.7 Every solvable set P has a solution S which is minimal relative to the (infinite) set of all finite blocks occurring in S.

Take any solution of P containing a subset K_1 (which may be P itself) of P. By 5.6, there is a subset M_1 of K_1 relative to which there is a minimal solution S_1 of P. Let K_2 be the set of all the 3 by 3 blocks occurring in S_1. Each member of K_2 is obviously made up of members of M_1. By 5.6, there is a subset M_2 of the union of M_1 and K_2 relative to which there is a minimal solution S_2 of P. Repeating, we can, for each n, define a set M_{n+1} and a solution S_{n+1}, minimal relative M_{n+1}, such that all blocks in S_{n+1} of size no greater than $2n+1$ by $2n+1$ occur in M_{n+1}. Treat now M_1, $M_2 - M_1$, $M_3 - M_2$, ... as giving partial solutions of P. We can now construct an infinite tree as in 5.1. By the infinity lemma, we obtain an infinite path which represents the solution S required by 5.7.

5.8 Given a domino set with a solution S, there is a solution T in which every finite block occurs infinitely often.

Consider, for $n=1$, 2, ..., all the $2n-1$ by $2n-1$ blocks which occur infinitely often in S. Clearly, for each n, there are only finitely many such blocks. Construct an infinite tree of these blocks as in 5.1. By the infinity lemma, we obtain an infinite path which represents the required solution T.

5.9 If a solution S is minimal relative to a set K of blocks, then every finite block occurring in S occurs infinitely often.

By the proof of 5.8, any block in K which occurs finitely often in S can be eliminated to get a new solution T. This would contradict the hypothesis that S is minimal relative to K.

5.10   Every solvable set has a solution S such that every finite block occurring in S occurs infinitely often in S.

This is direct from 5.7 and 5.9.

5.11   If a solvable set P has no periodic solutions, then it has as many distinct solutions as there are real numbers.

Since the set P is solvable, it has, by 5.10, a solution S in which every occurring finite block occurs infinitely often. Hence, if an n by n block occurs in S, it must have two nonoverlapping occurrences. Begin with two occurrences in S of a single tile T. There must be some n_1, such that the n_1 by n_1 blocks with T at the center at the two places are different. Otherwise, the two places are different. Otherwise, the two infinite columns C and D containing the two occurrences of T (or rows if they are in the same column) must be the same at corresponding positions, in which case we join the two occurrences of T by a staircase and consider all analogous staircases between the two columns. Since there are infinitely many staircases, at least two must be identical. But then we can take the region R bounded by C, D and two identical staircases and repeat it up and down to get an infinite strip Z bounded by the modified columns C' and D'. Since C and D are identical at corresponding places, the two vertical parts V_1 and V_2 bounding R are identical. Therefore, we can also repeat the infinite strip Z and cover the plane. Since each column consists of repetitions of V_1 (or, what is the same thing, V_2), there must be two infinite columns which are identical (i.e., without any staircase shift). Since there are infinitely many segments of rows bounded by the two columns, two of them must be identical. Therefore, we would have a torus, contradicting the hypothesis of nonperiodicity.

Hence, beginning with T and its two occurrences, we can expand to two different n_1 by n_1 blocks. Each n_1 by n_1 block has two nonoverlapping occurrences which, for similar reasons, can be extended to two distinct bigger blocks. Hence, we have, by repeating the process, a full binary tree with as many paths as there are real numbers. But each infinite path determines a solution.

5.12 Given an infinite set K of solutions of a domino set P, a solution S of P is said to be a limit solution of P and K if every finite block in S agrees with infinitely many solutions in K over that block.

Exercise 6. Every infinite set of solutions of a domino set has a limit solution; in other words, if there are infinitely many infinite paths, then there is a path on which every node appears in infinitely many infinite paths.

There is a solitaire called 15 puzzles which A. M. Turing once used in a popular article (in *Science news*) for the purpose of discussing algorithms. The game uses a 4 by 4 frame to hold 15 squares numbered 1 to 15. The small squares can be shifted by translation to a neighboring empty space. The interesting mathematical problem is to prove that one can never reach certain configurations.

APPENDIX B. ALGORITHMS AND MACHINES

1   Numerical and nonnumerical algorithms

In the game of nim, we have an algorithm (a general mechanical method) by which, given any three piles, we can decide whether the first or the second player has a winning strategy. This, we recall, can be done either by the laborous method of making a tree of all possible plays or by the shortcut using simple numerical calculations. Usually nonnumerical algorithms such as decision procedures in logic cannot be reduced to numerical ones in such a direct manner.

The most familiar mechanical procedures are those for the elementary arithmetic operations (the doing of "sums" in elementary schools). The essential components of computation are clear in these operations, if we reflect that we memorize the multiplication table only with products of all single-digit numbers taken two at a time, and do not attempt the impossible task of remembering directly products of all multi-digit numbers. Instead of an infinite table, a list of steps is memorized along side with the multiplication table to facilitate the handling of complicated multiplications. More generally, when we know how to perform certain simple operations, operations of greater complexity such as the extraction of square roots can be done according to fixed lists of steps. In school algebra, considerable time is spent on learning algorithms to solve single quadratic, cubic, and quartic equations, as well as simultaneous linear equations.

For the purpose of illustration, let us look more closely at the Euclidean algorithm for finding the greatest common divisor (g.c.d.) of two positive integers a and b. Since the g.c. d. of a and b is no greater than a or b, we could find their g.c.d. by brute force: try to divide a and b by 1, 2, ... in succession until we reach a or b. Then take the greatest number that does divide both a and b. The Euclidean algorithm is better in that, in general, it enables us to find the g.c.d. more quickly. For example, if $a = 153$, $b = 68$, we have: $153 = 2 \times 68 + 17$, $68 = 4 \times 17$, and the g.c.d. is 17. The usual description of the algorithm begins with the unnecessary assumption $a > b$ and continues as follows. Divide a by b, so that $a = q_1b + r_1$. If $r_1 = 0$, b is the g.c.d. Otherwise, divide b by r_1, getting $b = q_2r_1 + r_2$. If $r_2 = 0$, the g.c.d. is r_1. Repeating this process, we finally obtain $r_{n-1} = q_{n+1}r_n$, and r_n is the g.c.d.

Such a description is too informal for an ordinary computer to understand. To be more mechanical, we make use of labeled boxes (or more technically, registers with addresses) to store questions, answers, intermediate results, and instructions (steps to be taken; acts to be performed). This device not only avoids ambiguity but also makes it possible to be more explicit about repetitions (loops). Thus, for example, we may store a and b in registers A and B, the answer in register D, and intermediate results in registers C (as well as A and B). If we use $\langle x \rangle$ to represent the content of register x, an imagined program might look like this.

Address	Content	Remark
A	a	
B	b	
C		
D		
1	STO (REM (A, B), C)	Store remainder of $\langle A \rangle / \langle B \rangle$ in C.
2	TZE $(C, 6)$	Transfer to instruction 6 if $\langle C \rangle = 0$, otherwise continue with next instruction.
3	STO (B, A)	Store $\langle B \rangle$ in A.
4	STO (C, B)	Store $\langle C \rangle$ in B.
5	TRA (1)	Go to 1 (jump to 1, transfer to 1).
6	STO (B, D)	Store $\langle B \rangle$ in D.
7	PRINT (D)	Print out $\langle D \rangle$ as answer.

This defines a function $GCD(a, b) = \langle D \rangle$. If $a = b$, the instructions $1, 2, 6$ puts $\langle D \rangle = b$ as the g.c.d. of a and b. If $a < b$, rem $(a, b) = b$, instructions 3 and 4 interchange a and b. If $\langle A \rangle > \langle B \rangle$, the loop from 1 to 5 repeats divisions of $\langle A \rangle$ by $\langle B \rangle$ until a case is reached with remainder 0. Then instruction 2 exits to give the answer $\langle D \rangle = \langle B \rangle$.

Since storing $\langle x \rangle$ in y is the same as putting $\langle y \rangle = \langle x \rangle$, we can also rewrite the program in a somewhat different manner.

FUNCTION $GCD(M, N)$

(1) $K = \text{REM}(M,\ N)$

(2)   IF $K = 0$, GO TO 6

(3) $M = N$

(4) $N = K$

(5)   GO TO 1

(6) $GCD = N$

   END

Initially $\langle M \rangle = a$, $\langle N \rangle = b$, and the first instruction puts $\text{rem}(a, b)$ in K.

It is easy to imagine a machine which, given either of the above programs and the input a and b, will print out the g.c.d. of a and b as an answer. In fact, the second program is quite similar to what one would write in the languages of FORTRAN and ALGOL which are used in certain existing machines.

If the basic operation REM is not included in a given machine, one would normally expand the program to include a subprogram (subroutine) which will produce $\text{REM}(M, N)$ from M and N. Alternatively one may also avoid it by using repeated subtractions.

(1)   IF $M - N < 0$, EXCHANGE M and N

(2) $K = M - N$

(3)   IF $K = 0$, GO TO 9

(4) $M = K$

(5)   IF $M \geqslant N$, GO TO 2

(6) $M = N$

(7) $N = K$

(8)   GO TO 2

(9) $GCD = N$

There are a number of questions concerning the design and use of languages for writing programs, which we shall not discuss here.

A simple example of nonnumerical algorithms is the general method for finding a shortest path through a maze[1]. If we represent a maze by a (finite) connected graph with nodes joined by edges, then we can always find a shortest path in the following manner. First, suppose by shortest we mean simply that the number of nodes that one passes through is the smallest. Suppose we wish to go from A to B. First, we write the number 0 on node A. When we have written the numbers $0, 1, \ldots, n$ on all nodes which require these numbers, we write, if there remain nodes without a number, $n + 1$ on all nodes which have not yet been written on, but which are adjacent to (i.e., directly connected by an edge to) nodes which have. Since there are only finitely many connected nodes, all nodes will get numbers. The first time when node B gets a number, 6 say, we can trace a shortest path first to a node with number 5, then to one with number 4, and so on, until we reach A, which has the number 0. If we are also interested in the actual length of a path from A to B, then we write the numbers a little differently.

Exercise 1. Give an algorithm by which, given any finite connected graph (maze), the length of every edge, and two nodes A and B, we can find a shortest path from A to B.

The graph serves as a map of the maze. If, for example, like Theseus of the Greek mythology in search of Minotaur, we do not have a map of the maze, we have to travel back and forth between a node numbered n and one numbered $n + 1$, in order to carry out the above algorithm. Suppose the aim of Theseus is simply to find Minotaur by *some* path, a natural algorithm would be the following. Take an arbitrary path, marking every edge yellow on the way.

----

1) There is quite a bit of literature on efficient procedures for finding a shortest path through a maze and related questions, in, particular, the well-known one of the travelling salesman: to find a shortest path which passes through every node of a given finite connected graph at least once. For a solution of Exercise 7, see Edward F. Moore, "The shortest path through a maze", *Harvard Computation Laboratory Annals*, vol. 30 (1959), pp. 285—292. For a detailed discussion of Theseus in search of Minotaur, see B. A. Trakhtenbrot, *Algorithms and automatic computing machinery*, Boston, 1963, pp. 17—24.

Stop if Minotaur is found or if a dead-end is reached. In the second case, retrace the last edge to reach the neighboring node N, marking the twice-travelled edge red. Take an arbitrary path from N, avoiding only those edges which are marked by red but always following an unmarked path when there is one. When this process is repeated long enough, Theseus will certainly find Minotaur. Moreover, he can, by following yellow edges only, return to his friend Ariadne by a simple path, i.e., a path which passes no node more than once.

2   A programming prelude to abstract machines

We noted above the possibility of avoiding division by repeated subtractions in the Euclidean algorithm. More generally, one has standard methods of combining simpler instructions to obtain subroutines to perform more complex operations. Often we can find surprisingly small bases on which, in theory, we can build up programs for all computations. Since there are arbitrarily long numerical expressions, this possibility naturally depends on some idealization.

We consider first an idealized machine with an infinite sequence of registers numbered $1, 2, \ldots$, each of which can store an instruction or any nonnegative integer. Each particular program, however, involves only a finite number of these registers, the others remaining empty (i.e. containing 0) throughout any computation with the program. We enjoy considerable freedom in our choice of a set of basic instructions. The general principle is merely that in order to use the machine conveniently to carry out specific computations, we desire a large set of basic instructions; in order to prove general properties about a machine such as its limitations and the reducibility of it to other machines, we desire a small set of basic operations. These two desiderata can often be met simultaneously by showing that a larger set can be derived from (or reduced to) a smaller one.

As a first example, we consider a fairly small set of basic instructions. To shorten explanations, we use $\langle n \rangle$ and $\langle n' \rangle$ to denote respectively the content of register n before and after carrying out the instruction.

(a) $m + n = k$: add $\langle m \rangle$ and $\langle n \rangle$ and store sum in register k, i.e., $\langle k' \rangle = \langle m \rangle + \langle n \rangle$.

(b) $m - n = k$: subtract $\langle n \rangle$ from $\langle m \rangle$ and store difference in register k, i.e., $\langle k' \rangle = \langle m \rangle - \langle n \rangle$.

(c) $S(1, n)$: store the number 1 in register n, i.e., $\langle n' \rangle = 1$.

(d) $J(m, n; k)$: compare $\langle m \rangle$ and $\langle n \rangle$, and jump to instruction k if $\langle m \rangle \leqslant \langle n \rangle$.

(e) $\bar{J}(m; k)$: jump to instruction k, if $\langle m \rangle = 0$.

(f) $C(m, n)$: copy from register m into register n, i.e., $\langle n' \rangle = \langle m \rangle$.

A finite sequence of instructions of the above forms make up a program. The next instruction in order is followed, unless otherwise specified in a jump instruction (type d or e). In this language, the last program for the Euclidean algorithm can be rewritten thus.

2.1   A program of the Euclidean algorithm with the above base (a and c not used).

(1)	a	the given integers are a and b.
(2)	b	
(3)		space for intermediate data and final answer.
(4)	$J(2, 1; 8)$	if $b \leqslant a$, jump to instruction 8, otherwise
(5)	$C(1, 3)$	exchange $\langle 1 \rangle$ and $\langle 2 \rangle$.
(6)	$C(2, 1)$	
(7)	$C(3, 2)$	
(8)	$1 - 2 = 3$	put $\langle 1 \rangle - \langle 2 \rangle$ in register 3
(9)	$\bar{J}(3, 15)$	go to instruction 15 if $\langle 3 \rangle = 0$.
(10)	$C(3, 1)$	put $\langle 3 \rangle$ in register 1.
(11)	$J(2, 1; 8)$	if $\langle 2 \rangle \leqslant \langle 1 \rangle$, go to instruction 8.
(12)	$C(2, 1)$	put $\langle 2 \rangle$ in register 1.
(13)	$C(3, 2)$	put $\langle 3 \rangle$ in register 2.
(14)	$J(2, 1; 8)$	jump to instruction 8, since $\langle 2 \rangle \leqslant \langle 1 \rangle$.
(15)	$C(2, 3)$	put $\langle 2 \rangle$ in register 3, which gives the g.c.d. of a and b.

If we are interested in the economy of basic instructions, we can simplify the above set consisting of (a) to (f) considerably. If, for convenience, we add an auxiliary register numbered 0 (or ∞), we can prove that for each program from the above set of instructions, we can find a corresponding program with the same effects on the following machine.

2.2   The unlimited registered machine (URM). It has an infinite sequence of registers numbered $0, 1, 2, \ldots$, each of which can store an instruction or any nonnegative integer. The basic instructions are of three types:

(α) $P(n)$: add 1 to the number in register n, i.e., $\langle n' \rangle = \langle n \rangle + 1$.

(β) $D(n)$: subtract 1 from the number in register n, i.e., $\langle n' \rangle = \langle n \rangle - 1$. (this instruction is used only when $\langle n \rangle \neq 0$).

(γ) $J(n; k)$: jump to instruction k if $\langle n \rangle \neq 0$.

In order to prove the adequacy of (α), (β), (γ) to represent programs with (a) to (f), we use subroutines to represent other instructions.

2.2.1   Subroutine for unconditional jump, say a line j. $J(k)$.
j. $P(0)$, $J(0; k)$.

2.2.2   Suppose the line is: j. $\bar{J}(m; k)$. Replace it by:
j. $J(m; j+1)$, $J(k)$.

2.2.3   Suppose the line is: j. $C(n)$, i.e., clear register n or put $\langle n \rangle = 0$. Replace it by: j. $J(n; j+1)$, $D(n)$, $J(j)$.

2.2.4   If I is an instruction or a subroutine, I^n stands for the result of performing I n times. Suppose the line is: j. $I^{\langle n \rangle}$. Replace it by: j. $\bar{J}(n; j+1)$, I, $D(n)$, $J(j)$.

2.2.5   Suppose the line is: j. $C(m, n)$. Replace it by:
j. $C(n)$, $C(0)$, $\{P(0), P(n)\}^{\langle m \rangle}$, $\{P(m)\}^{\langle 0 \rangle}$.

This is a rather complex replacement because the result becomes several lines and the rest of the program has to be renumbered accordingly. The same is true of 2.2.6, 2.2.7, 2.2.9 below.

2.2.6   Suppose the line is: j. $m + n = k$. Replace it by:
j. $C(n, 0)$, $C(m, k)$, $\{P(k)\}^{\langle 0 \rangle}$.

2.2.7   Suppose the line is: j. $m - n = k$. Replace it by:
j. $C(n, 0)$, $C(m, k)$, $\{D(k)\}^{\langle 0 \rangle}$.

2.2.8   Suppose the line is: j. $S(1, n)$. Replace it by:
j. $C(n)$, $P(n)$.

2.2.9   Suppose the line is: j. $J(m, n; k)$. Replace it by:
j. $C(m, 0)$, $\{D(0), \bar{J}(0; j+1)\}^{\langle 0 \rangle}$, $J(k)$.

The above sketch of how to obtain (a) to (f) from (α), (β), (γ) is expected to convince the reader that the URM contains an adequate set of basic instructions. The following exercise should assist in making one's understanding more definite.

Exercise 2. Rewrite the program 2.1 by using only instructions of the forms (α), (β), (γ). If abbreviations are used, give the number of instructions in the whole program in terms of the basic instructions of the URM.

When doing this exercise, one would probably find it highly desirable to have a device of calling a subroutine in such a way that the calculation continues after the subroutine is performed. Since the URM does not have explicitly such provisions, a same subroutine has to be copied again each time it is applied. Many existing computers have ways of keeping a record of the next instruction to be performed after calling in a subroutine. Incidentally, we can also obtain (α), (β), (γ) from (a) to (f).

So far, we have assumed that the registers of the URM may store not only numbers but also instructions. This means that we permit the possibility of modifying the program as the computation proceeds. Since, however, we do not actually take advantage of this facility, we may also think of the instructions as stored in a separate unit. In other words, the registers store only the numbers. In fact, for certain purposes, this interpretation is more convenient.

Under this interpretation, we can view the registers as, instead of a storage medium, a finite sequence $\langle 0 \rangle$, $\langle 1 \rangle$, ..., $\langle n_t \rangle$ of integers at each moment t. If we write each nonnegative integer n by a string of n 1's, we may think of $\langle 0 \rangle$, $\langle 1 \rangle$, ..., $\langle n_t \rangle$ as a single word A on the alphabet $\{1,c\}$, c for the comma. A particularly elegant formulation is possible which can be shown to imitate the URM faithfully.

2.3 *The* SS *machine*[1]. The alphabet is $\{1, c\}$. The programs operate on words (finite strings) A over the alphabet, and are made up of instructions of the following types.

(A) Pl: add 1 to the end of A.

(B) Pc: add c to the end of A.

(C) D: delete the first symbol of A.

(D) $J(k, m)$: jump to instuction k or m according as the first

1) This and related machines are considered in J. C. Shepherdson and H. E. Sturgis, ''Computability of recursive functions'', *Journal ACM*, vol. 10 (1963), pp. 217—255.

symbol in A is 1 or c.

To simulate a program on the URM, we represent the registers at the beginning by the word $A = \langle 0 \rangle e \langle 1 \rangle c \ldots c \langle N \rangle$, in which N is the least number such that all registers which are either nonempty initially or mentioned in the program have number $\leqslant N$. To simulate the three operations (α), (β), (γ), we need a device to bring the content $\langle n \rangle$ of the relevant register n to the beginning of the word A, perform the operation, and restore the word A to its original position.

This is done by repeated uses of the following subroutine:

2.3.1 T: 1. Pc, $J(2,3)$, 2. D, Pl, $j(2,3)$, 3. D.
This changes $N_1cN_2c \ldots cN_p$ to $N_2c \ldots cN_pcN_1$.

The instructions $P(n)$, $D(n)$, $J(n;k)$ can be replaced as follows.

2.3.2   Replace j. $P(n)$by j. T^{n+1}, Pl, T^{N-n-1}.

2.3.3   Replace j. $D(n)$ by j. T^n, D, T^{N-n}.

2.3.4   Replace $j.J(n;k)$by: j. T^n, $J\left(j + \dfrac{1}{3}, j + \dfrac{2}{3}\right)$. $j + \dfrac{1}{3}$.

$$T^{N-n}, J(k;k), j + \frac{2}{3}. \quad T^{N-n}.$$

It is assumed that $j + \dfrac{2}{3}$ is followed by $j + 1$. Actually, the renumbering would be more complex than indicated because, by 2.3.1, T also requires renumbering.

The reduction of the URM to the SS machine illustrates the possibility of examining one symbol, instead of one number, at a time.

3   Human computation and practical computers

There are a number of basic questions centering around the very notion of a mechanical process. A great achievement of logic since the 1930's is the success in giving an absolute (i.e., not depending on the particular formalism chosen) definition of the interesting epistemological notion of mechanical processes (effective procedures, computability, algorithms, finite methods). In fact, this may be said to be the only basic epistemological concept relating to mathematics that we have been able to illuminate with fairly complete success so far. Other concepts on the same level, viz provability, definability, and conceivability, have been treated exactly with some success only relative to certain special formal systems.

This success was brought about, more than anything else, by Turing's mathematical formulation of a type of abstract machines,[1] based on a philosophical analysis of the process of human computation. Our belief in the correctness in Turing's analysis has been further strengthened by the independent development of the interesting equivalent concepts of (general and partial) recursiveness and Post's combinatorial production systems, and by the accumulated experience that, as far as we know, all intuitively computable functions are indeed Turing computable.

That such a definitive analysis is possible is very surprising. Given such an analysis, there remains the vital problem of relating theory to practice, finite experience to infinite potentiality, and the problem of recognizing a proposed algorithm to be a genuine algorithm. In particular, although the analysis gives a precise answer to the question of what machines in theory can or cannot do, it leaves wide open what human activities can feasibly be taken over by machines in the reasonably near future.

An algorithm is a finite set of rules which tell us, from moment to moment, precisely what to do with regard to a given class of problems. Given an algorithm for solving a class K of problems and a problem belonging to *K*, anybody can solve the problem provided he is able to perform the operations required by the algorithm and to follow exactly the rules as given. For example, a schoolboy can learn the Euclidean algorithm correctly without knowing why it gives the desired answers. In practice, when a man calculates, he also designs small algorithms on the way. But to simplify matters, we may say that the designing activity is no longer a part of his calculating activity.

Thus, in following an algorithm, a human calculator uses cer-

1) A. M. Turing, ''On computable numbers'', *Proc. London Math. Soc.*, vol. 42 (1936), pp. 230—265 and vol. 43 (1937), pp. 544—546. The paper is reprinted in Martin Davis, *The undecidable*, 1965.

tain data, including the instructions of the algorithm, which he remembers in part and are mostly written down on sheets of paper or printed in reference tables or charts. In short, there is (a) some storage device to store initial data and intermediate results. In addition, he performs (b) certain elementary operations such as multiplication or integration. Moreover, he exercises (c) control of the sequencing of steps by consulting the instructions to determine what step is to be performed next.

Of these three components, (b) is the easiest to mechanize, at least when operations are sufficiently elementary. The introduction of desk calculators delegates a considerable portion of this part to machines. What the automatic (sequencing) computer does is to add (1) a memory unit to store information, (3) a control unit to carry out the instructions according to a program which embodies the algorithm. In addition, the desk calculator is replaced by (2) an arithmetic unit, which is integrated with (1) and (3). At first the program was given externally (by plugboards) so that the machine cannot be instructed to modify the program to add flexibility. But soon machines began to include "stored programs" in such a way that the program is treated as part of the initial data and, with ingenuity, one program can be written to instruct the machine to do automatically the work of a bundle of programs.

This sketch emphasizes numerical calculations. But it is known that nonnumerical algorithms can often be realized on existing machines since, for example, letters of the English alphabet can be represented by numbers.

The above description gives in the roughest terms an outline of the concepts behind existing computers. It leaves out on the one hand the enormous engineering research and development needed in building and improving the actual computer, and on the other hand a careful analysis of the basic notion of computations and algorithms.

4   A conceptual analysis of computations[1]

The intuitive notion of an algorithm is rather vague. For

1) For related general discussions, *compare From mathematics to philosophy*, 1974, pp. 81—99 and Chapter X.

example, what is a rule? We would like the rules to be mechanically interpretable, i.e., such that a machine can understand the rule (instruction) and carry it out. In other words, we need to specify a language for describing algorithms which is general enough to describe all mechanical procedures and yet simple enough to be interpreted by a machine. Thus, instead of the one problem of defining an algorithm, we are led to two equally difficult problems of defining a mechanical language and a machine. We seem to be moving in circles. One might suggest that we should collect a large sample of algorithms and analyse them to find out their common feature. But it is hard to see how such inductive approach in the manner of J. S. Mill can, without some appropriate powerful ideas, produce a satisfactory abstraction.

What Turing did was to analyse the human calculating act and arrive at a number of simple operations which are obviously mechanical in nature and yet can be shown to be capable of being combined to perform arbitrarily complex mechanical operations. Qualitative increase in the complexity of an algorithm is substituted by quantitative increase in memory size and time needed for carrying out the algorithm. Moreover, once such a simple conception of a machine is given, the problem of designing a mechanical language also gets solved easily.

If we think of a man computing on a piece of paper, which we may suppose is divided into squares, we find the following things all involved in the process. (1) A storage medium, viz. the piece of paper; (2) a language, with symbols to represent numbers and directions which, for simplicity we may assume, are written down on the piece of paper; (3) scanned regions, i.e., certain squares are observed at each moment; (4) "states of mind," viz, at each stage the man keeps track of the stage of computation and decides what step is to be taken next; (5) the act of taking the next step of computation which may involve (a) a change in the symbols by writing or erasing (crossing out) certain symbols, (b) a change of the scanned region, (c) a change of the "state of mind."

What makes the process mechanical, i.e., capable of being performed by a machine, or by a man in a mechanical (noncreative) way, can be summarized in the following two principles.

4.1   *The principle of determinacy.*

4.2   *The principle of finiteness.*

According to the first principle, the only relevant information in deciding what act is to be performed next is the symbols currently observed in the region under scan and the present "state of mind." Now in computing according to an explicit algorithm, the algorithm is supposed to specify what is to be done under all possible circumstances arising from the computation. But the only way the present circumstances at each moment can be delineated for the man doing the computation is through his present experience, viz. the symbols (on the paper) currently observed plus his present "state of mind."

The rather vague notion "state of mind" calls for some further examination. In a way, the symbols currently observed, as observed, is part of the man's present experience, but isolating them as a separate unit makes the situation more precise. The present "state of mind" may contain many irrelevant factors, human nature being what it is. The relevant factors which need to be taken into consideration is primarily the mind's response to the symbols currently observed in the light of the instructions of the algorithm and the process of computation as carried out thus far. Now, in the normal course of a human computation, the memory of what one has done thus far may help to decide the next step to be taken. However, we believe that if one is sufficiently careful, this indeterminate element can be made exact and written down on the piece of paper. In this way we may think of the present "state of mind" as an exact conditional attitude of being ready to perform certain specific acts depending only on what the currently observed symbols on the piece of paper are. And this agrees with our concept of a precise algorithm which specifies that at each stage different specific acts are to be performed according to the data currently available. The "state of mind" is merely to keep track of what the present stage is in the whole process of computation.

According to the principle of finiteness, the mind is only capable of storing and perceiving a finite number of different items at each moment; in fact, there is some fixed finite upper bound on such items. The upper bound on the number of items perceived at each moment is rather small, while that on those stored would be quite large although it still seems reasonable to agree that there is some such upper bound. From this principle, it follows immediately that at each moment only finitely many squares can be scanned. If the computing individual wishes to observe more, he must use successive observations.

Moreover, the number of states of mind which need be taken into account is also finite, because these states must be somehow stored in the mind, in order that they can all be ready to be entered upon. An alternative way of defending this application of the principle of finiteness is to remark that since the brain as a physical object is finite, to store infinitely many different states, some of the physical phenomena which represent them must be "arbitrarily" close to each other and similar to each other in structure. These items would require an infinite discerning power, contrary to the fundamental physical principles of today. In fact, there is a limit to the amount of information that can be recovered from any physical system of finite size.

As a third application of 4.2, we may assume that the number of symbols which can be printed is finite. An infinity of different symbols cannot be stored in the human mind, even though we seem to feel we can, in theory, make arbitrarily long combinations of a certain small number of given symbols. We distinguish single symbols from compound symbols by the criterion that we must be able to observe a single symbol at one glance. Given a finite bound to the area we can observe at a glance, an infinity of symbols must include ones which differ to an arbitrarily small extent, and to recognize them as different would require capacity to make arbitrarily minute distinctions. In any case, to make proper use of an infinity of symbols would require states of mind which are infinitely complex. We can argue against the assumption of a single state of mind which is infinitely complex, with much the same arguments as those used against the supposition of infinite number of states of mind.

The fourth application of 4.2 is to the operations performed at each step of the computation. As we noted before, the operations involve three types of changes: the state of mind, the symbols as written on the working sheet, and the region under observation. It may seem rather idle and frivolous to introduce a state and then change it without changing the written symbols or the region under observation. But, since the operation is determined by the state and the symbols observed, the same state may cause other changes when other symbols are observed. In fact, even a state which always causes only changes of states is useful because the difference in the symbols observed may effect a conditional transfer to different states. For much the same reason, it is possible to keep the state unchanged while making a change of another type. We may, therefore, think of each operation as either a change of just one of the three types, or a change of both the state and one of the two other factors, or a change of all three items. Depending on which alternative is chosen, one obtains different but equivalent formulations of the abstract machines. The choice, however, makes no difference for our immediate purpose here, and we shall for the moment assume that each operation makes only one of the three types of change.

Since there are only a fixed finite number N of states in each problem of computation, there are no more than $N(N-1)$ possible changes of states. Similarly, since there is a fixed finite bound B to the number of squares scanned at each moment and there are only finitely many different symbols, there are a finite number of distinct observable regions, and finitely many possible changes of such regions. In view of this fact, we could either represent each possible region (e.g., n^2 connected squares on the piece of paper with "blank" counted as a symbol) by a single symbol so that each time only a single symbol is observed and changed; or else, we may avoid the increase of the total number of available symbols but observe or change only a single symbol at a time. Other observations and changes can be split into simple ones of this kind. This simplification would, in general, entail an increase in the number of states. But that is all right.

It remains to consider the change from observing one region on the piece of paper to observing another. The new region must be immediately recognizable by the computing individual. It would seem reasonable to suppose that the distance between the two successively observed regions does not exceed a f x amount, say L units. If we assume, as we just defended, that each time a single square is scanned, then the situation is slightly simplified since we can easily delineate the possible units next to be observed in a region of $(2L + 1)^2$ squares with the currently scanned square at the center. As before, we may reduce L to 1 by splitting a move to unit steps, possibly with the cost of increasing the number of states.

One relatively minor point is the tacit assumption of the two-dimensional nature of the storage (and working) space, viz. the piece of paper. Using a one-dimensional medium would simplify somewhat the description of the abstract machines and make no theoretical difference. As we grow up, we tend to abandon the two-dimensional computation we did in our childhood, and we are convinced that a one-dimensional paper or tape, when long enough, is adequate. Moreover, as the one-dimensional model is developed further, it will be clear that a higher dimensional medium does not increase the range of what is computable in theory, because any computation with that medium can be simulated by one on a one-dimensional tape.

We may now construct a mechanical model of the computation with a given algorithm. Each state of mind corresponds to a "configuration" or state of the machine. The machine scans B squares at each moment. In any operation, the machine can change its configuration, can change a symbol on a scanned square, or can change the scanned region to another region no more than L squares away. As we have observed, there is no loss of generality if we restrict both B and L to be unity and let the machine work on a linear tape. Hence, the only variable elements are the number of available distinct symbols (the size of the alphabet) and the number of states. Since with a larger alphabet we can embody more information in a given state, it is possible to reduce the number of states by enlarging the alphabet and reduce the size of the alphabet by increasing the number of states.

These abstract models are what are known as Turing machines which will be treated more formally in the Appendix C.

Turing's basic thesis is:

4.3 *Computability = Turing computability.* Any procedure is effective (can be embodied in an algorithm) if and only if it can be realized by a Turing machine.

5   Five contrasts concerning machines

5.1   *Nature and artifacts.*

In analysing human computation to arrive at an elucidation of the concept of computability, much of the argument would be clearer if we think in terms of building a machine (an artifact) to do the computation, instead of the natural phenomenon of computations by mind.   For example, it seems rather straightforward to say that we can only build machines which have a finite number of states and use a finite alphabet.   To argue that the mind also has only such finite capacities in doing computations, we had to offer more elaborate reasons which dovetail with difficult philosophical problems pertaining to the concept of mind.   And it is not easy to eradicate the vague feeling that perhaps mind can do certain computations which machines can never do.

If we agree that we are only interested in computation as a mechanical process, we can avoid philosophical problems of psychology such as the relation of mind and body.   But there remain problems of space and time.   For example, since time is infinite, it is possible that a result (e.g. a real number) computed by eclipses of the moon, or by proving new theorems, cannot be obtained by any Turing machine.   Or since space is infinite and matter is infinitely divisible, the physical world might contain sequences of 0's and 1's not computable by Turing machines.   Should one admit that such real numbers or sequences are computable?   The answer is probably that we do not feel an effective procedure or algorithm is thereby given for the computation.   More pertinently perhaps, it seems far-fetched to bring in the proposition that the physical world is infinite.

These questions may also be avoided by insisting that we are concerned with artifacts rather than nature.   Even though it is possible to make machines which "grow", the growth must be of such a type that the blueprint contains all the germs.   And then, once the pattern of growth is formalized, we expect that what such machines can compute can also be computed by Turing machines.

In the formulation of Turing machines, we do envisage a potentially infinite tape so that we may say that the tape grows longer as more squares are needed. This potential infinity is, however, perfectly orderly and presents no serious new conceptual problem relative to our mathematical knowledge. It is basically no harder to understand than the proposition that there are infinitely many integers. The problems in juxtaposing infinity with machine are rather those concerned with the virtues of alternative idealizations in order to obtain a significant (and preferably applicable) theory of machines. We shall soon return to this topic.

5.2 *Theoretical possibility and practical feasibility.*

In the formulation of Turing machines and the analysis leading to it, we have completely neglected the element of speed or efficiency. We merely argued that any class of computations which a machine can do, some Turing machine can do it too. We did not claim that the Turing machine can do it as quickly or as efficiently. In fact, the last proposition is eminently untrue. Turing machines are notoriously slow.

This question of efficiency presents an entirely different set of problems. For example, both addition and multiplication can be computed by Turing machines, and it takes longer to do the multiplication. However, in order to establish satisfactorily that multiplication is a more complex type of computation than addition, it is not enough to consider Turing machines only. Rather we have to think of all sorts of abstract models and show that in every case, on the whole multiplication takes longer than addition. This whole area of efficiency and complexity of computations is of much current research interest and probably very difficult.

A more abstract aspect of the question of efficiency is connected with the contrast between existence and the method of finding the object. It is customary to think of a Turing computable function of nonnegative integers as one for which there is a Turing machine such that if the arguement values are represented initially on the tape, the machine will eventually halt with a representation of the function value on the tape. It can be shown that the Turing computable total functions are exactly the general recursive functions. Now, given a Turing machine, in order that it does define a computable total function in the above sense, it must be true that for every nonnegative integer m (as input), there exists a number k of steps such that the machine will halt in less than k steps (and leave the function value as the "output"). But the proof of such existence statements can be very difficult. In fact, for any sufficiently rich formal system S, it is possible to give some such statement that is undecidable in S.

This problem about existence is not serious if one accepts a classical point of view, since we assume a definite notion of the totality of nonnegative integers. Moreover, if we do not separate out the total functions from all the Turing computable ones, we can bypass the existence statement. A deeper reason for the feeling that a Turing computable function might not be effectively computable is the fact that, in general, we do not have an estimate, for given argument values, of how long it will take before the process stops or whether it will even stop. In addition, to decide a class of problems usually requires a total function that will terminate for each question of the class.

While these considerations do not prove that Turing computability is too broad a notion, they do point to the desirability of closer studies on the complexity of computations with a view to introducing certain informative hierarchies within the whole range of Turing computable functions.

Another aspect of the contrast between theory and practice is the limitations of computers. On the one hand, "a machine can only do what it is told", i.e., any procedure that can be performed by a computer can be precisely described. This, however, does not mean that the designer of a computer program cannot be surprised by correct but unexpected results produced by the computer. Many computer programs, perhaps too many of them in the current practice, are frankly experimental, i.e., to find out the behavior under diverse circumstances of a specified method or system, which is not fully understood in advance. The situation here is to some extent similar to the familiar phenomenon of surprising new theorems deducible in some perfectly specific formal system.

On the other hand, the analysis of the preceding section leads to the heuristic principle that any procedure which can be precisely described can be programmed to be performed by a computer. Hence, it would seem that all we need, in order to have machines accomplish clever feats, is to find a precise description of the procedure by which they are done.   In practice, our understanding of psychology and physiology is far too imperfect to justify a belief that in the foreseeable future we can faithfully duplicate major feats by such simulations.   In practice, the more successful computer efforts are often done by designing mechanical procedures different from what is natural to people.   And then the question of practical feasibility ("the explosion of combinatories") presents strict demands on avoiding crude mechanical methods.   It is in this sense that the wish to extend areas of interesting computer applications calls for the study of a new type of mathematics that is interested more in practical feasibility than theoretical possibility.   Because it is less easy to get elegant results about practical feasibility, there is a basic conflict in that the human mind cannot feasibly handle a great multiplicity of complex and inelegant procedures.   To be optimistic, we may remind ourselves that conflicts are a precondition of progress.

5.3   *Programs and states.*

In analysing the process of computation, we have emphasized states (of mind).   A more familiar framework is to think in terms of programs consisting of individual steps.   Apart from the states, the individual acts are: move one square along the tape, write or erase.   If, as is possible, we use only an alphabet with two symbols, one of which is taken as the blank, then there are just four acts: move left one square, move right one square, mark, erase.   Now we can give instructions to have the machine perform these three kinds of act in numerous different orders.   But this is not adequate for at least two reasons.   We use only finite lists of instructions so that each list only tells the machine to go through less than a fixed number of squares.   But in general, we want it to be possible that the numbers of squares which are to be operated on can increase indefinitely even with a fixed set of instructions.   A more obviously serious defect is that a method must enable us to give different answers to different questions, and therefore some element of recognition and choice must be present in the instructions. All these defects can be removed by adding an additional type of instruction called conditional transfer in computing machines actually in use.   By this type of instruction, the machine is to recognise whether the square under scan is blank   and   then   follow   two   different   instructions according as whether it is blank or not.

In this way we arrive at a single computer on which each Turing machine is represented by a program. The program is not much different from the description of the states of the Turing machine. It does have the advantage of distinguishing routine operations from the conceptually more complex operation of transfers. From the point of view of a designer of computers, it is more natural to think in terms of states, while from that of a user of computers, it is more natural to think in terms of programs.

5.4 *Use versus design*.

The different orientations of the user and the designer present certain practical problems. The more flexible a programming language is, the easier it is for a programmer to write a program for a given problem, and the harder it is for a designer to design a machine realizing such a language in the hardware. Often it is so hard to include the desired flexibility in the hardware, that one resorts to the designing of compilers, i.e. programs which permit the use of more flexible languages by translating them into the more rudimentary language available through the hardware of the machine.

The analysis of computations in the last section was centered around the act of computation. We proceed now to examine the simplifying assumptions which yield a natural model of machines as physical objects.

We may think of an arbitrary physical object (a rock, a machine, an animal, a solar system) that changes its state in time, i.e. it changes (e.g. through rearrangment of elementary particles). It may or may not change its size in time, it may or may not interact with its environment. We shall make an abstract consideration of such a physical object treated as an idealized machine: a logical, or symbolic, or informational structure. First we neglect all problems of physics (and engineering) such as the physical composition and the transition of energy and power. We shall also leave out the past influence of the environment, except insofar as it is recorded in the present state of the object. The further assumptions are:

5.4.1 *Discrete units*. We use only discrete descriptions. Between any two moments or two elements, there are only finitely many other moments or elements.

5.4.2 *Deterministic behavior*. The present state of the object (the machine) is entirely determined by its present environment (current input) and past history, including the effects of its environment through the past.

5.4.3 *Finitude of bases*. Each system will contain a finite number of elements at each time and each element can be in only one of a finite number of states at any time.

Under these assumptions, the state at a time $t + 1$ is determined by a fixed relation R:

$$S(t + 1) = R[I(t), S(t)].$$

Moreover, since at each time the machine is only capable of finitely many states, there is no point in assuming the possibility of an infinite past: even if there had been an infinite past, $S(t)$ can only distinguish them as finitely many distinct equivalent classes. This question is actually immaterial for our present discussion, since we get as a corollary from 5.4.1—5.4.3:

5 4.4 The state at $t + 1$ is determined by the state and the input at t.

Of course we have not made the assumption that the machine is altogether only capable of a fixed finite number of states. It is not excluded that the machine might "grow" in some predetermined manner. For example, in a Turing machine that only operates on a finite tape at each moment, we may think of endmarks on the tape and an instruction that adds a new tape square when an attempt is made to shift beyond one of the two endmarks. According to our present framework, a Turing machine is then capable of infinitely many states because we make no separation here between an internal state and the tape configuration, but both are viewed as parts of the present state.

If we do make the assumption that the machine is not capable of growth, then we get what is known as sequential machines or finite automata, which are a particularly simple mathematical structure. Elevators, combination clocks, switches, all may be taken as examples of such machines. For many purposes, it is also convenient to think of a Turing machine as consisting of a finite automaton interacting with a potentially infinite tape. It is not quite right to view the tape as the environment because it forms an integral part of the Turing machine so that it is largely under the control of the finite automaton part and it influences the latter in a predetermined manner.

We have considered only total states of a machine and avoided the mention of the local actions involved in the interactions of different parts. These interactions are of course essential for the actual building of computers. For example, it is desirable to control these interactions by assuming synchronous operations. And a certain simple logic (Boolean algebra) finds applications in detailed studies about building blocks.

5.5 *Finite and infinite state machines.*

Mathematically Turing machines are more significant than finite automata. But since we can only make finite machines, why should we study the infinite model? There is of course the ready answer that the physicist, even though he can only measure and observe finite things, uses the calculus and studies classical analysis. But there is more. It is unfair to ask a finite automaton questions about arbitrarily large numbers. However, our whole mathematical thinking has been molded in such a way that, at least in theoretical considerations, we are generally accustomed to thinking about all positive integers. As a result, any fixed finite bound tends to introduce unnatural restrictions and make it torturous when we try to ask sharp general questions or to get sharp general answers. For example, a finite automaton cannot even multiply two arbitrarily large numbers. The fact is, our intuitive conception of computations requires a more comprehensive framework than finite automata.

Even though the simple formulation of Turing machines is not a realistic model of existing computers, it seems reasonable to suppose that, in general, infinite state machines are, compared with finite ones, a more useful guide to the general considerations, even practical ones, about computers. The practical limitations rarely take the form of the lack of an adequate amount of machinery. In fact, for instance, the possible supply of external storage tapes could be said to be potentially infinite. The trouble is rather that the access time is much too slow. In another direction, the number of possible states of existing computers are much too large for any finite-state analysis to be of much use in a global way.

If it is not entirely clear whether finite or infinite state machines will turn out to be more fruitful for the future use and design of computers, it is relatively easy to agree that there is, for the purpose of the theory and practice of computers, no point in studying actually infinite models, viz., those violating 5.4.3 or having an infinite speed of effective operation.

APPENDIX C.   ABSTRACT MACHINES

1   Finite-state machines

1.1   *Transducers, recognizers, and logical nets.*

The ordinary electric switch has the two states "on" and "off", while there are many more complex switches from the four-state control of a three-way light to twelve-state switches on some television sets.   Other familiar examples of finite-state machines include combination locks and elevators.   To most of us who are users rather than designers, these function as black boxes with input and output signals. For an abstract treatment, we shall ignore the interpretation of these signals but merely single out the fact that each channel is capable of some fixed finite number of states, which, without loss of generality, will be taken to be 2.

Our basic assumption about the black box is that each machine consists of a finite set of discrete parts, such as cores and transistors, each of which is capable of a finite number (this may be, without loss of generality, taken to be 2) of states.   As a corollary of this assumption, each black box is only capable of finitely many internal states.

A finite-state machine máy be studied either in terms of the total states or in terms of its internal structure, i.e. how different parts are put together to realize the various states.   From the first point of view, we can regard the machine either as a transducer or as a recognizer.   From the second point of view, the machine is regarded abstractly as a logical net.

  1.1.1   A sequential machine (finite-state transducer) is a 6-tuple $M = (Q, I, O, g, h, q_0)$ where ($Q(t)$ being the state at time t)

(i) Q is a finite set of states,

(ii) I is a finite set of inputs,

(iii) O is a finite set of outputs,

(iv) g is a function from $Q \times I$ into Q: $Q(t+1) = g[Q(t), I(t)]$,

(v) h is a function from $Q \times I$ into O: $O(t+1) = h[Q(t), I(t)]$,

(vi) q_0 is a distinguished member of Q: $Q(0) = q_0$.

Often we may omit q_0 and (vi), either because we are not interested in the initial state or because we have a natural choice of the initial state, such as the one with all parts quiescent. If there are m input channels and n output channels, then I and O have 2^m and 2^n possible members respectively, since each channel is capable of 2 states. Similarly, if the machine has k basic "parts," there are at most 2^k possible members in Q.

1.1.2 *Sequential machines as finite-state recognizers.*

The recognition problem may be viewed as a special case of interpreting the output signals. For this purpose, we assume a single output channel which emits either the signal 0 or the signal 1 at each time. For each t, the input history up to t is said to be accepted by the machine if and only if the output at $t + 1$ is 1. In this way, the input histories are divided into two classes, say the good ones and bad ones. For each class K of input strings, we may ask whether a machine recognizes it, i.e., emits the signal 1 at $t + 1$ if and only if the input history up to t belongs to K.

1.1.3 *Logical nets*

A basic circuit is of one of the following four forms:

An or-circuit emits an output 1 if and only if at least one input channel carries the signal 1; otherwise the output is 0. An and-circuit emits 1 if and only if both inputs are 1. A not-circuit emits 1 if and only if the input is 0. A d-circuit delays one time unit. In this formulation, the first three units are supposed to be instantaneous. Alternatively, we may assume a built-in delay of one time unit in each of the first three elements and delete the delay circuit as a separate primitive unit. In that case, we shall use circles instead of triangles.

A logical net is defined inductively thus. (1) A basic circuit is a logical net. (2) If A and B are two logical nets with output channels x and y, then

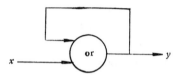

are logical nets. In particular, we permit branchings of the same channel to serve as inputs to different logical nets; we permit also disconnected parts.

1.2   *Examples.*

1·2.1   *An existence machine.*   We consider the following simple net:

This net remembers whether there was ever a time when the input x was alive, i.e. carried the input signal 1.

Construed as a recognizer, it accepts every finite string of 0's and 1's which contains at least one 1.

In terms of states, we have the ignorant state q_0 and the learned state q_1.

$Q(0) = q_0$.

$Q(t + 1) = q_0$ and $O(t + 1) = 0$, if $I(t) = 0$ and $Q(t) = q_0$.

$Q(t + 1) = q_1$ and $O(t + 1) = 1$, if $I(t) = 1$ or $Q(t) = q_1$.

1.2.2   *The serial binary adder.*   Two input channels carry simultaneously two strings of binary digits with least significant digits first. A single output channel yields the binary digits of the sum of the two numbers fed into the machine. Hence, $O = \{0, 1\}$, $I = \{00, 01, 10, 11\}$. There are only two states: a "carry" state q_1 and a "no-carry" state q_0. It can be verified that:

$Q(0) = q_0$.

$Q(t + 1) = q_0$ if $I(t) = 00$ or 01 or 10, and $Q(t) = q_0$;
        or $I(t) = 00$ and $Q(t) = q_1$.

$Q(t + 1) = q_1$ if $I(t) = 01$ or 10 or 11, and $Q(t) = q_1$;
        or $I(t) = 11$ and $Q(t) = q_0$.

$$O(t + 1) = 0 \text{ if } I(t) = 00 \text{ or } 11, \text{ and } Q(t) = q_0;$$
$$\text{or } I(t) = 01 \text{ or } 10, \text{ and } Q(t) = q_1.$$
$$O(t + 1) = 1 \text{ if } I(t) = 00 \text{ or } 11, \text{ and } Q(t) = q_1;$$
$$\text{or } I(t) = 01 \text{ or } 10, \text{ and } Q(t) = q_0.$$

1.2.3 *The parallel binary adder.* The parallel binary adder for getting the sum of two numbers with n binary digits each uses $2n$ inputs and $n + 1$ outputs. It is obvious that we cannot design a parallel adder to add two numbers of arbitrarily many digits since we would not have enough input channels. But using the idea of the serial adder, we can, in theory, design a sequential machine to multiply an arbitrary number by a fixed number n. For instance, we may connect up $(n - 1)$ serial adders to add any number m to itself $(n - 1)$ times.

1.2.4 *Counters up to any fixed number.*

1.3 *Limitations of finite machines.*

It should be noted that if the input function assumes a fixed constant value throughout or if there is no input at all, then $Q(t + 1)$ would become a function of $Q(t)$ only and the machine would eventually either half or repeat itself. Left to itself, a sequential machine, once in periodic pattern, can never get out of it. Hence, without appropriate assistance from the input, a sequential machine can never output an infinite sequence which is not ultimately periodic. Moreover, the period cannot have length more than the number of states.

1.3.1 *Squaring a number and multiplication.*

We mentioned before the possibility of multiplying an arbitrary number by a fixed number n. If, however, we wish to find a machine which can multiple an arbitrary number by itself, then the problem is quite different. For a sequential machine with k states, we can always find a number n such that n^2 has too many more digits than n. For example, if n has m digits, n^2 has more than $m + 2k$ digits. But once the input stops, the machine is left to its own devices and can only begin to repeat itself after printing out at most $m + k$ digits. For example, if we assume the channels to be capable of ten states each and use the decimal notation, 10^{2k} squared would yield 10^{4k} with $4k + 1$ digits. After the input stops and we have obtained $2k + 1$ digits of output (all 0), the k-state machine can print the $2k$ more required 0's only if it prints 0 ever after, thereby failing to put out the digit 1 as the $(4k+1)$-th digit. Hence, no finite-state machine can do arbitrary squaring or multiplication. Other tasks beyond the capacity of such machines include:

1.3.2   *Unlimited counter.*

1.3.3   *Parenthesis matching.*

1.3.4   *Decoding parenthesis-free notation.*

1.4   *The firing squad and the expensive negator.*

1.4.1   *The firing squad synchronization problem.*

Consider a finite (but arbitrarily long) one-dimensional array of finite-state machines, all of which are alike except the ones at each end. The machines are called soldiers, and one of the end machines is called a General. The machines are synchronous, and the state of each machine at time $t+1$ depends on the states of itself and of its two neighbors at time t. The problem is to specify the states and transitions of the soldiers in such a way that the General can cause them to go into one particular terminal state (i.e., they fire their guns) all at exactly the same time. At the beginning (i.e., $t=0$) all the soldiers are assumed to be in a single state, the quiescent state. When the General undergoes the transition into the state labeled "fire when ready," he does not take any initiative afterwards, and the rest is up to the soldiers. The signal can propagate down the line no faster than one soldier per unit of time, and their problem is how to get all coordinated and in rhythm. The tricky part of the problem is that the same kind of soldier with a fixed number K of states is required to be able to do this, regardless of the length n of the firing squad. In particular, the soldier with m states should work correctly, even when n is much larger than m. Roughly speaking, none of the soldiers is permitted to count up as high as n.

Outline of a solution. Suppose first there are $n=2^k$ soldiers. We successively find the middle of the string of soldiers and start over, with twice as many squads as before, each having half the previous number of soldiers. When a soldier discerns that he is both the left and the right end of a squad, he fires, and this will happen to all soldiers simultaneously. To find the middle of a

nonnegative integers, and since we have a better developed theory of numbers, it is customary to concentrate our attention to computable functions from numbers to numbers.

If we wish to study numerical computations by Turing machines, a first problem is choose a notation to represent numbers, i.e., a way of using certain words as numerals. There is room for different choices which are convenient for different purposes. Let us assume that the alphabet is $\{0, *\}$ so that each tape square is capable of only two states: blank or marked. Let us use the notation a^n to represent that the symbol or string a is repeated n times.

For example, if we wish to multiply two positive integers m and n, we may choose the input and output "formats" (a term familiar in the use of digital computers) in various ways. Examples are:

(i)   input: $0 *^m 0 *^n 0$; output: $0 *^m 0 *^n 0 *^{m \cdot n} 0$.

(ii) input: $0 *^{m+1} 0 *^{n+1} 0$; output: $0 *^{mn+1} 0$.

(iii)   input: $0(*0)^{m+1} 00(*0)^{n+1} 0$; output: $0(*0)^{mn+1}0$.

(iv)   input: $0(*0)^{m+1} 00(*0)^{n+1}0$;
         output: $00(*0)^{m+1} 00(*0)^{n+1} 000(*0)^{mn+1}0$.

The format (i) is not suitable if we consider nonnegative integers rather than positive integers because we would need a special notation for the number 0 to avoid the ambiguity of not distinguishing it from blank. In the second format, each number n is represented by $*^{n+1}$ so that, in particular, $.*$ is the notation from the number 0. For our immediate purpose, we shall choose the format (ii), which can clearly be generalized to take care of functions with more than two arguments. Formats (iii) and (iv) provide more "scratch paper."

2.3.1   A function $f(x_1, \ldots, x_n)$ is said to be Turing computable, or briefly computable, if there exists a Turing machine T which computes f. And T_f is said to compute f in case that, for any n-tuples $m_1, \ldots, m_n$ $f(m_1, \ldots, m_n) = k$ if and only if, beginning with the initial tape

$$\overset{\downarrow}{0} *^{m_1+1} 0 \cdots 0 *^{m_n+1} 0 \quad (\downarrow \text{ indicates square under scan}),$$

T_f will halt with the tape containing

$$\overset{\downarrow}{0} *^{k+1} 0.$$

We shall use the notation:

$$(x_1, \cdots, x_n)_{T_f} \to f(x_1, \cdots, x_n)$$

to represent that T_f computes f.

The above definition does not assume that f is total, i.e. it is defined for all argument values. In case f is a partial function, T_f computes f provided the values agree over the domain on which f is defined. For the other argument values, T_f must never halt. We shall adopt the terminology that f is a partial computable function if f is a partial function and it is computable. The briefer term "computable function" is reserved for total functions which are computable. Hence, every computable function is partial computable but not conversely.

As a simple example, we show that addition is computable.

2.3.2 $f(x, y) = x + y$.

We wish to find T_f such that

$$\overset{\downarrow}{0} *^{m+1} 0 *^{n+1} 0$$

is transformed to:

$$\overset{\downarrow}{0} *^{m+n+1} 0.$$

We simply have the machine head go right along the tape to find the first blank square and mark it; continue right till the first blank square, erase two symbols on its left; and go left to stop at the first blank square. The machine T_f is given by:

	0	1
1	→, 2	
2	*, 3	→, 2
3	←, 4	→, 3
4	←, 5	e, 4
5	←, 6	e, 5
6		←, 6

Observe that no action is specified when T_f is in q_1 scanning a marked square, or in q_6 scanning a blank. The convention is, under such cases, the machine halts. It is customary to avoid in this way an additional notation for "stop".

With a little effort, one can also find Turing machines which compute $f(x, y) = xy$, $f(x) = x!$, $f(x, y) = x \doteq y$ (i.e., $x - y$ if $x \geqslant y$ and 0 otherwise), and so on.

2.4 *Equivalent formulations of Turing machines.*

The capabilities of Turing machines as a whole class are quite stable. We can change various factors and the resulting machines will retain the same capacity. We shall list a few well-known possibilities. But we shall not supply any proofs since these results are rather remote from our central interests. We shall not even specify the exact sense in which two classes K_1 and K_2 are equivalent, except to say that simple correlations between computations in K_1 and K_2 can be set up so that for any machine in either class we can find one in the other with essentially the same computing capacity.

In particular, the class of computable (numerical) functions for each formulation remains the same.

2.4.1 *One-way and two-way infinite tapes.*

2.4.2 *Two-symbols and m-symbols* $(m > 2)$.

2.4.3 *Multiple Tape machines.*

2.4.4 *Multi dimensional storage units.*

2.4.5 *Quadruples, quintuples, triples.*

The formulation above uses quadruples in the sense that we specify for each state and each scanned symbol, both a single act (shift or print) and the next state. In the quintuple formulation, we specify each time a shifting act followed by a printing act. In the triple formulation, we specify only one "output" which is either a change of state or a shift act or a print act.

It is known that, by using enough symbols, each Turing machine can be reduced to one with two states in the quintuple formulation. Recently it has been established that this reduction to two states is not always possible if the quadruples are used instead.

The quintuples are normally taken to be of the form $q_a s_i p_j d q_b$, where p_j is print s_j and d is $\leftarrow$ or $\rightarrow$. It would not do to use a formulation that reverses the order of print and shift, because in that case the effects of inputs, with the possible exception of the content of the initially scanned square, are eliminated. Thus, since each quintuple takes the form $q_a s_i d p_j q_b$, after the initial moment, only symbols printed by the machine are observed. For example, no such quintuple machine can do addition or multiplication.

3   The *P*-machine (a program formulation of Turing machines)

There are many differences between Turing machines and modern digital computers. Among these are the use of instructions instead of state-symbol combinations and the fact that a Turing machine realizes a single algorithm while on a computer various programs can be written to realize diverse algorithms. The *P*-machine to be introduced behaves like a digital computer in these two respects but remains in other ways more similar to Turing machines. This model lies in a middle ground where basic features of complex computers are illustrated but the primitive apparatus is stripped to a bare minimum.

3.1   *Specification of the P-machine.*

3.1.1   The imagined machine *P* is made up of four parts:

(i)   An indefinitely expandable internal random-access (parallel) storage to store programs (each program is a finite sequence of instructions).

(ii)   An indefinitely expandable linear tape (a serial storage), rolled into a succession of squares, to store input, output and intermediate results. For definiteness, we assume that each square is only capable of two states: blank (contains 0) or marked (contains 1).

(iii)   A control element which keeps track of the program step to be taken at each moment.

(iv)   A read-write head which, at each moment, is supposed to be scanning one and only one square on the tape, and which is capable of shifting one square (left or right), reading, and writing (marking or erasing).

At each moment, the next operation of the machine *P* is determined by the instruction of the program under attention of the control element, together with the content (blank or marked) of the square under scan. The machine *P* can perform five kinds of operations in accordance with these two factors: (a) the read-write head moves one square to right (by the way, the same purpose can be accomplished by shifting the tape one square to the left); (b) it moves one square to the left; (c) it marks (i.e., prints 1 in) the square under scan; (d) it erases (i.e. prints 0 in) the square under scan; (e) the control element shifts its attention to some other instruction in the program: it "decides" to follow one of two preassigned instructions according as whether the square under scan is marked or blank (frequently the decision is immaterial and the next instruction in the program is followed unconditionally).

To store the program in the internal storage, we shall assume that there are registers with addresses which can be any positive integer. Each register is capable of storing one instruction and can be referred to by using its address.

3.1.2 *Basic instructions.* Corresponding to the five types of operations, there are five types of basic instructions:

(1) $\rightarrow$ shift the head one square to the right.

(2) $\leftarrow$ shift left.

(3) $*$ mark the square under scan.

(4) e erase the square under scan.

(5) $C(n)$ a conditional transfer.

Of these (1)—(4) are four single instructions while the conditional transfer really embodies an infinite bundle of instructions. When these instructions occur in a program, they get addresses. A conditional transfer takes on the form $m.C(n)$, according to which the m-th instruction word (i.e., the instruction word in the register with address m) is a conditional transfer such that if the square under scan is marked, then follow the n-th instruction: otherwise, i.e., if the square under scan is blank, follow the next instruction, i.e., the $(m + 1)$-st instruction.

3.1.3 A program (or routine) on the machine P is a set of ordered pairs such that there exists a positive integer k such that (α) the set consists of k pairs; (β) for every n, n occurs as the first member of exactly one pair in the set if and only if $1 \leqslant n \leqslant k$; (γ) the second member of each pair is either $\rightarrow$ or $\leftarrow$ or $*$ or e or $C(n)$, with $1 \leqslant n \leqslant k$.

For example, the program

1. $\rightarrow$

2. $C(1)$

causes the head to move right along the tape until a 0 is found.

The program

1.   *, 2. →, 3. →, 4. *, 5. $C(2)$.

produces the sequence 010101... on a blank tape.

The convention is implicit that, unless specified otherwise by a conditional transfer, the next instruction is followed. When $n = m + 1$, the instruction $m.C(n)$ is wasteful, although it need not be excluded.

There is nothing in the definition 3.1.3 to prevent the occurrence of an instruction to mark a marked square or erase a blank square, even though we can usually so construct the programs that when operating on inputs which interest us no such wasted actions will arise. To preclude, for example, the marking of marked squares in the general definition of programs would require unnecessary complications: one could replace every step $m.$* by two steps $m. C(m + 2)$, $(m + 1).$*, and renumber all steps $m + i$ in the original program as $m + i + 1$.

Since there is no separate instruction for halt, it is understood that the machine will stop when it arrives at a stage when the program contains no instruction telling the machine what to do next. For uniformity and explicitness, one could agree to end each program with two instructions such as: $k - 1.$ →, $k.$ ←.

3.2   *Equivalence of turing machines with programs on the P-machine.*

It is quite obvious that given a (two-symbol) Turing machine, we can find a P-program to do the same thing; and conversely.

The proof is omitted. The reader may supply the proof, using as illustration a program and a Turing machine for, say, addition or multiplication or a conversion routine from unary to binary notations of integers.

3.3   *Macro-instructions and subroutines.*

It is not necessary to write out all the labels for the instructions in a program. The labels are needed only if the instruction is referred to elsewhere in the program. For example, the two simple programs in 3.1.3 can be rewritten as:

1.   →, $C(1)$.

1.   *, 2. →², *, $C(2)$.

One example would be to find n^2:

$$(1)^{n+1} \xrightarrow[P(n^2)]{} (1)^{n^2+1}$$

Assume the program starts with the first 1 under scan and ends with the last 1 under scan.

Other examples would be:

(1) m^n.

(2) Copy a consecutive string of 1's or an arbitrary block.

(3) Decide whether a string of 1's represents a prime number.

(4) Check whether a string of parentheses is well-formed.

(5) Find a program which enumerates all numbers in the binary notation, with a blank between two consecutive numbers.

(6) Enumerate all well-formed formulas defined by: p is well-formed, if A and B are well-formed, so is CAB. (The basic alphabet is $\{p, C\}$).

(7) Conversion between parenthesis and parenthesis-free notations for the above class.

(8) Find a program which, beginning with a blank tape, will halt with more 1's on the tape than 10 times the number of basic instructions in the program.

(9) Find a program which, beginning with a blank tape, will make more shifts ($\rightarrow$ or $\leftarrow$) than 10 times the number of basic instructions in the program.

Often it is more convenient to use additional symbols beyond 0 and 1.

Consider a program which produces 0010110111011110... from a blank tape:

RTZ: 1. $\rightarrow$, $C(1)$. (a macro-instruction).

LTZ: 1. $\leftarrow$, $C(1)$.

Program: 1. $\rightarrow^2$,

2. $e, (RTZ)^2, *, (LTZ)^2, *, \leftarrow, C(2), (RTZ)^2, *, C(2)$.

An open subroutine, when used, is essentially inserted in place of a reference to it, while a closed subroutine permits a suitable calling sequence to transfer control to the subroutine and then return, without making a copy of the subroutine. A macro-instruction is more or less like a short open subroutine, except that it permits more flexibility such as changes in the arguments and applications in printing required tables.

3.4   *A universal P-program (a universal Turing machine).*

Each *P*-program or Turing machine computes a function from (input) words to (output) words. It can also be interpreted, by suitable representations, as computing various functions. In view of the fact that every *P*-program has only finitely many instructions we can enumerate them or represent them by numbers and words. Hence, we may think of each *P*-program as an argument of a function and find, with some effort, a universal *P*-program *U* such that, for each program *P* if *p* represents *P* and

$$W \xrightarrow[P]{} f(W),$$

then

$$p, W \xrightarrow[U]{} f(W).$$

The construction is fairly straightforward, especially if we allow additional symbols. Since a 2-symbol formulation is equivalent to a formulation with more symbols, we begin the exposition by using a larger alphabet. Actually, we could code the tape words and the programs and their operations in various ways. Any exact enough description of a program *P* would enable us to trace out the behavior of *P,* and the universal program could work as an "interpretive" programming system.

For simplicity, we shall assume that for *P* we operate on a one-way infinite tape and the programs always begin by scanning the beginning of the tape. For *U,* we use a two-way infinite tape but use a letter *L* to divide the tape into two halves. The alphabet consists of the following symbols:

$$0, 1, 2, 3, 4, 5, 6, 7, 8, 9, B, M, \cdot, *, \rightarrow, \leftarrow, e, C, ;, S, L, T$$

To simulate a program *P,* we simply copy the program at the beginning of the tape without abbreviation (except for omitting some of the dots and the parentheses in conditional transfers) followed by the given tape word. The additional sympols *B, M, S* are for indications about the current instruction and the currently scanned symbol: *S* after an instruction (replacing ;) indicates the current instruction,

B indicates a currently scanned 0, M indicates a currently scanned 1. For example, if P is

$$1.\ C(3);\quad 2.\ \rightarrow;\quad 3.\ *;\quad 4.\ C(2)$$

and the initial tape word is 1101, then the initial tape for the program U is:

The program U can be described by using in addition to $\rightarrow$ and $\leftarrow$, the basic instructions $C(s, n)$, $\bar{C}(s, n)$, $W(s)$, for each basic symbol s so that, in particular, $C(1,\ n)$ corresponds to $C(n)$, $W(0)$ and $W(1)$ correspond to e and $*$. $\bar{C}(s, n)$ transfers to n if the symbol is not s.

1. $\rightarrow$, $\bar{C}(;, 1)$, $W(S)$ Find first instruction and mark by S.
2. $\rightarrow$, $\bar{C}(\cdot, 1)$, $\rightarrow$, $C(1, 3)$, $W(B)$, $C(B, 4)$
3. $W(M)$ Find first tape symbol and write B or M.
4. $\leftarrow$, $\bar{C}(S, 4)$
5. $\leftarrow$, $C(*, 7)$, $C(e, 8)$, $C(\rightarrow, 9)$, $C(\leftarrow,)$
6. $\leftarrow$, $\bar{C}(C, 6)$, $C(C, 25)$

At this stage, we compare the numbers before and after C in order to determine whether the next instruction is to be sought on the left or on the right of the current instruction. We also use the portion of the tape to the left of L to find out the difference of two numbers. Once this is accomplished, we can count off the; symbol with the auxiliary symbol T and find the instruction to be transferred to. Then we print; over S, S over T and continue with instruction 5 given above.

7. $\rightarrow$, $C(M, 17)$, $\bar{C}(B, 7)$, $W(M)$, $C(M, 17)$ mark
8. $\rightarrow$, $C(B, 17)$, $\bar{C}(M, 8)$, $W(B)$, $C(B, 17)$ erase
9. $\rightarrow$, $C(B, 10)$, $\bar{C}(M, 9)$, $W(1)$, $C(1, 11)$ shift right
10. $W(0)$
11. $\rightarrow$, $C(1, 12)$, $W(B)$, $C(B, 17)$
12. $W(M)$, $C(M, 17)$
13. $\rightarrow$, $C(B, 14)$, $\bar{C}(M, 13)$, $W(1)$, $C(1, 15)$ shift left
14. $W(0)$

15. $\leftarrow$, $C(\cdot, 19)$, $C(1, 16)$, $W(B)$, $C(B, 17)$

16. $W(M)$

17. $\leftarrow$, $\bar{C}(S, 17)$, $W(;)$, $\rightarrow$, $C(\cdot, 19)$ go to next instruction

18. $\rightarrow$, $\bar{C}(;18)$, $W(S)$, $C(S, 5)$

19. Stop

Exercise. Write an easily understandable universal program either by working out the above sketch or by devising a better one.

In order to write a universal P-program U, we use auxiliary squares to help avoid auxiliary symbols. To be specific, we use every other square as an auxiliary square. We code the instructions by:

$\rightarrow$	11
$\leftarrow$	1111
*	111111
e	11111111
$C(n)$	$2(n + 4)$ 1's.

If, for example, the program is

$$1. \rightarrow, \quad 2.\ C(1)$$

and the initial tape is 1101, we represent these on the initial tape for U by:

$$0001\overset{\downarrow}{1}101\underbrace{11111111111}_{C(1)}0010\underbrace{110111}_{\text{data}}000\cdots$$
$$\underbrace{}_{\rightarrow}$$

We use two simple macro-instructions:

$$RTZ:\ 1. \rightarrow^2, C(1). \qquad LTZ:\ 1. \leftarrow^2, C(1)$$

The program U is as follows:

1. $\rightarrow^2$, $C(2)$, $\rightarrow$, $C(11)$, $\leftarrow^2$, $C(16)$ $\rightarrow$ or stop
2. $\rightarrow^2$, $C(3)$, $\rightarrow$, $C(12)$, $\leftarrow^2$, $C(16)$ $\leftarrow$ or stop
3. $\rightarrow^2$, $C(4)$, $\rightarrow$, $C(14)$, $\leftarrow^2$, $C(16)$ * or stop
4. $\rightarrow^2$, $C(5)$, $\rightarrow$, $C(15)$, $\leftarrow^2$, $C(16)$ e or stop
5. $\rightarrow$, e, $\leftarrow$ transfer

6. $\rightarrow^2, C(5), \rightarrow, C(7), RTZ, \leftarrow, C(8),$
    $LTZ, C(16)$ decide whether last instruction

7. $RTZ, \leftarrow, C(8), \rightarrow, (LTZ)^2, *, \leftarrow,$
    $RTZ, \rightarrow^2, C(1)$ not last instruction but scan 0

8. $(LTZ)^2, \rightarrow^4,$ scan 1

9. $e, RTZ, *, \rightarrow, C(4), \rightarrow, LTZ, *,$
    $LTZ, \rightarrow^2, C(1)$ find new instruction

10. $\rightarrow, e, LTZ, *, \leftarrow, RTZ, \rightarrow^3, C(9)$

11. $\leftarrow, e, (RTZ)^2, *, \rightarrow^2, e, (LTZ)^2,$
    $*, \rightarrow^3, C(1)$ shift right

12. $\leftarrow, e, (RTZ)^2, *, \leftarrow^2, C(13), C(16)$ left shift

13. $e, (LTZ)^2, *, \rightarrow^3, C(1)$ if not off tape end

14. $\leftarrow, e, (RTZ)^2, \leftarrow, *, \rightarrow, (LTZ)^2, *,$
    $\rightarrow^3, C(1)$ mark

15. $\leftarrow, e, (RTZ)^2, \leftarrow, e, \rightarrow, (LTZ)^2, *,$
    $\rightarrow^3, C(1)$ erase

16. $\leftarrow, \rightarrow.$ stop

The above program is not elegant and may contain mistakes. Correct, elaborate, reorganize or otherwise improve it.

3.5 *Nonerasing machine and recursive functions.*

A surprising discovery about the program machine is that erasing can be avoided altogether. While this is only of theoretical interest as far as we know, it is conceivable that some attractive physical device might be discovered for which it is expensive or slow to erase (i.e., it is easy to change the state in one direction but hard to reverse it).

Such a machine B is obtained from the machine P by deleting the erase operation e. In the machine, a symbol, once written, will never be changed. In fact, since we use only two symbols, the machine B can only change blank squares on its tapes to 1's, but cannot change a 1 back to a blank.

We can establish the universality of the machine B by proving that every recursive function is B-computable. Since it is easy to show that every P-computable function of numbers is recursive, this proves the adequacy of the B machine for dealing with functions of numbers. It is also possible to prove either analogously or by coding words as numbers that this is true for functions of words as well. We shall leave out these developments[1] and bring in familiar concepts of recursion theory in terms of machines and programs.

We shall assume as given that a (partial) function is recursive if and only if it is computable (by a P-program or a Turing machine).

3.5.1   A set is recursive if its characteristic function is general recursive. The characteristic function f of a set S is given by:

$$f(n) = 0 \iff n \in S$$
$$f(n) = 1 \iff n \notin S.$$

A set S is recursively enumerable (r.e.) if it is empty or there is a general recursive function f such that:

$$n \in S \iff \exists m(f(m) = n).$$

In terms of programs, we obtain immediately:

3.5.2   A set is recursive if there is a program which prints out 0 for inputs belonging to the set and 1 for inputs not belonging to the set. A set is r.e. if there is a program which prints out each member of the set and no other numbers, as the inputs range over numbers.

There are some easy theorems about recursive and r.e. sets.

3.5.3   Every recursive set is r.e.

If P is the program for a recursive set S, write P_1 which copies the input tape at a safe place (which is possible), performs P, and then either keep the argument value if the result is 0 or else change the argument value to a number in S. If S is empty, S is also r. e. by definition.

------

1)   The program machines P and B were introduced in *Journal ACM*, vol. 4 (1957), pp. 53—92. These are sometimes called Wang machines in the literature. According to M. Minsky, ''The first formulation of Turing machine theory in terms of computer-like models appears in the paper Wang 1957, which contains results that would have been much more difficult to express in the older formalisms,'' *Computation: finite and infinite machines*, 1967, p. 200.   See also a different treatment of nonerasing machines on p. 262.

3.5.4   If a set is recursive, so is its complement.

Just modify the program to exchange 0 and 1.

All these proofs can be made more exact when definite conventions are made explicit on the computations by a program.

3.5.5   If a set and its complement are each r.e., the set is recursive

Given the functions f and g enumerating S and S', we can get a program to do the following on the input n: compute $f(0)$, $g(0)$, $f(1)$, $g(1)$, etc. and compare the result with n each time. Continue until a result n is obtained, which is always possible since n is in the range of f or g. If we get n while computing f, print 0; otherwise, print 1. The actual program is easier with some auxiliary symbols.

3.5.6   A set is recursive if and only if both it and its complement are r.e.

3.5.7   By 3.5.3, 3.5.4, 3.5.5, a set S is r.e. if and only if there is a program which, beginning with a blank tape, will "print out" each member of the set and no other members. For example, we may leave a gap of several blanks, perform computations on the right and record answers on the left.

To prove this by 3.5.2, we first assume a program P_1 such that $P_1(0)$, $P_1(1)$, ... together give the set S. Then we can define a program P_2 from P_1 so that P_2 generates 0, computes $P_1(0)$, records it, generates 1, computes $P_1(1)$, etc.

We note that the universal program or machine considered in 3.4 contains the idea of "stored program computers" which enables us to store programs (coded into machine words, i.e., integers) in the machine in the same manner as data. This has the advantage of modifying given programs more quickly. Moreover, it makes possible the creation of operating systems (OS, or executive programs) which can digest incoming object programs written in more sophisticated languages and carry out the instructions in terms of more primitive machine operations.

If we think about the universal program, it is easy to see that we can enumerate or encode as natural numbers all possible programs. Moreover, the entire configuration (the instruction number, the tape content, the square under scan) at any instant of carrying out a program can be coded in a number. And a computation record for a program is but a finite sequence of such numbers from the initial state to the halting state. The whole idea of a program or a computer implies that the state at $t + 1$ is determined in a simple manner by the state at t. Hence the computation record of a program P on an input x can be coded in a number again provided the machine eventually halts beginning with P and x.

Putting together these observations and taking only functions of one argument for brevity, we can convince ourselves that there is a recursive ternary relation T which holds of e, x, and y provided e codes a program P, and y is the computation record of the universal program beginning with e and x. Moreover, once the computation terminates, we can recover from y in a mechanical way the answer on the tape. In other words, there is a recursive function U such that whenever $T(e, x, y)$ holds, $U(y)$ is the output of computing from e and x, i.e., $U(y) = \phi_e(x)$. In other words, $\phi_e(x) = U[\text{the least } y \text{ such that } T(e, x, y)]$. The number e is usually called the index of the function ϕ_e. A partial function is recursive iff it has an index. A function ϕ_e is general recursive iff $\forall x \exists y \ T(e, x, y)$. These conclusions are known under the name of "normal form theorem". It is possible to have e range over all natural numbers by an enumeration of all programs of the machine P.

3.5.8   There is a (primitive) recursive function U and a (primitive) recursive ternary relation T such that, for every positive k, $\phi_e^{(k)} (x_1, \ldots, x_k) = U[\mu y T(e, \langle x_1, \ldots, x_k \rangle, y)]$; in other words, every partial recursive function can be expressed in terms of U and T in a uniform manner.

3.6   *Nonrecursive functions and unsolvable problems.*

Once we have a simple uniform way of dealing with all Turing machines or all (definitions of) partial recursive functions, it is easy to define functions which are not recursive. For example, let $f(n) = \phi_n(n) + 1$. This cannot be a general recursive function because for each general recursive function ϕ_e, $f(e) > \phi_e(e)$. From this it follows that:

3.6.1.   Neither $\{(x, y) \mid \phi_x(y) < \infty\}$ nor $\{x \mid \phi_x(x) < \infty\}$ is a recursive set.

This is usually called the unsolvability of the halting problem. But it may be more instructive to think in terms of computer programs especially for those who are more interested in computers.

3.6.2 *The universal halting problem.* There is no P-program H such that for every P-program P and every finite input tape word W, with descriptions p and w, $H(p, w)$ will always halt and yields an answer 0 or 1 according as P halts on W or not.

When a program, with input tape, is started in operation, it may take a long time before the computation is completed and the machine comes to a halt. The computation may go on forever, i.e., for certain program-tape pairs, it will never halt. And it is natural to ask whether we have a decision procedure to tell us the 'yes' or 'no' answer in each case. It will not do to take the universal program as giving the procedure since, although if P halts on W, we shall eventually know, we shall never find out merely by operating U persistently that P will never halt on W.

In fact, 3.6.2 says there can be no such program at all. Suppose there were such a program H, it would, in particular, also decide for the special case $w = p$:

$$p, p \xrightarrow[H]{} \begin{cases} 0 (\text{i.e., } *) \text{ if } P \text{ halts on } p, \\ 1 \ (\text{i.e., } **) \text{ if } P \text{ not.} \end{cases}$$

Let H_1 be a simple copying program so that

$$p \xrightarrow[H_1]{} p, p.$$

Combining H_1 and H, we get a program H_2:

$$p \xrightarrow[H_2]{} \begin{cases} * \ \ \text{ if } P \text{ halts on } p. \\ ** \text{ if } P \text{ does not halt on } p. \end{cases}$$

In particular, H_2 can be so chosen as to halt while scanning the (first) $*$. Suppose H_2 has q lines. Let H_3 be obtained from H_2 by adding the following:

$$(q + 1). \quad \rightarrow, C(q + 2), \leftarrow, C(q + 1).$$
$$(q + 2). \quad \rightarrow, \leftarrow$$

Hence, we have:

$$p \xrightarrow[H_3]{} \begin{cases} \text{never halts if } P \text{ halts on } p. \\ ** \text{ if } P \text{ does not halt on } p. \end{cases}$$

Since this is true for every program P, it is true when we take H_3 as P. But then we arrive at a contradiction.

It may be of interest to prove this theorem in terms of actual computers rather than abstract machines. The following proof was communicated to me by Yang Chen-Ning. Since he does not specialize in logic, his proof is likely to be more transparent to a wider group of people.

Consider a computer with infinite memory.   Any speed.

A *program P* is a deck of cards.

A *data D* is a deck of cards.

A *run* (P, D) or (P, D, D') is the loading of (P, D) or (P, D, D') into machine and starting it. If nonsensical the machine will stop. Comma is a "separater" card.

Lemma.   There does not exist a program A_1, with the property that when (A_1, X) is run, where X is any deck of cards, the machine always stops, leaving a

> printout "stops", if run (X, X) stops,
> printout "does not stop", if run (X, X) does not stop.

Proof.   If A_1 existed, we can slightly modify its cards near the exit to create A_2 with the property that when (A_2, X) is run, where X is any deck of cards, the machine

> does not stop if run (X, X) stops,
> stops if run (X, X) does not stop.

Then if (A_2, A_2) stops it should not stop.

If (A_2, A_2) does not stop it should stop.

Contradiction.

Theorem.   There does not exist a program A_3 with the property that when (A_3, X, Y) is run with any decks X and Y the machine always stops leaving a

> printout "stops" if fun (X, Y) stops
> printout "does not stop" if run (X, Y) does not stop.

Proof.   If A_3 exists, we can modify it at the beginning to form A_4, so that in running (A_4, X), the machine first duplicates X and then runs (A_3, X, X). A_4 is then the program A_1 of the Lemma, which does not exist.

One can write a program P to check successively all quadruples of positive integers to test whether they satisfy the Fermat relation. And it follows that P halts if and only if. Fermat's conjecture is false. This illustrates how difficult even an individual halting problem could be. The same applies to Goldbach's conjecture although not to the question whether there are infinitely twin primes.

$$\forall x\, \exists y\, (x < y \text{ and } y \text{ and } y + 2 \text{ both are primes}).$$

From the above discussion, we have the following result:

There exists no computable function $f(p, w)$ such that for every program P and every tape word W (with descriptions p, w) $f(p, w)$ is an upper bound on how long the computation of P with input W can take if it will eventually stop.

Thus, if there were such a function, we could test for each pair P and W whether the computation stops before $f(p, w)$ operations.

If now we consider the number of operations which any non-halting n-line program can perform with input n, there must be a maximum since there are only finitely many such programs for each n. Hence, we have a function $f(n)$ which gives such maxima. For the same reason, $f(n)$ is not computable. By argument similar to that used for the next statement, the initial tape can also be assumed blank.

3.6.3 *Halting problem on a blank tape.* No program can decide whether any program will halt with the initial tape blank.

For each program with an input word, we can find another program which, beginning with the blank tape, will produce the input word and then behave like the original program.

Using a universal program, we can also get:

3.6.4 *Halting problem for a single program.* There is no program which can decide whether a particular universal program halts for a given input word. (The machine is fixed, the inputs may very.)

3.6.5 *The printing problem.* No program can decide, for each program, whether a particular symbol s is ever printed by the program starting from a blank tape.

3.6.6 *The uniform halting problem.* No program can decide, for each program whether it halts on all input words.

One writes a new program that will erase all the material on the given tape. This would be true even if we abolish the assumption that only finitely many squares are nonblank, because a program has a fixed initial instruction. The erasing is done in stages so that the ground is always cleared in advance.

The results on these unsolvable problems can be proved analogously for any general definition of algorithms. Indeed, one can also consider all algorithms expressible in the English language, say, as used today.

3.6.7   There is an r.e. set which is not recursive (equivalently, whose complement is not r.e.).

For example, the set $\{x|\phi_x(x)<\infty\}$ is not recursive as mentioned before. But it is r.e. because it is the same as $\{x|\exists\, yT(x,x,y)\}$. Another example is the set of n such that P_n (in a given enumeration of programs) halts on a blank tape.

In terms of these notions, the requirement on axiom systems being entirely formal is often referred to as recursively axiomatizable. The commonly accepted stipulation is:

3.6.8   The set of proofs in any formal system is recursive; the set of theorems is r.e.

Hence, 3.6.7 gives an absolute incompleteness result in the following sense. We look for a complete theory for the set S as given in 3.6.7, i.e., for each n, $n\varepsilon s$ or $n\notin S$ should be provable according as whether n belongs to S or not. 3.6.7 says no axiom system can do this. Thus, if a system F does this, then, since the set of theorems is r.e. and we can decide whether a theorem is of the form $n\notin S$, the complement of S would be r.e., which is impossible.

This does not give any particular example of an incomplete system; it also assumes implicitly that numbers and formulas can be regarded in the same way. A particular system might have no means of representing S (or, in general, all r.e. sets), or contain no negation, or be inconsistent. Moreover, the above argument assumes the intuitive concept of truth.

4   Unsolvable tiling problems

We have just given some basic examples of unsolvable problems (non-recursive sets) in terms of Turing machines and their program counterparts. Many other mathematical problems which appear unrelated to these mentioned have been shown to be unsolvable by arguing that their solution would lead to a solution of, say, the halting problem. We shall give an example here in terms of some tiling problems on the basis of the discussions in Appendix A. These tiling problems can also be described in a different manner.

Assume given a finite set $P = \{D_1, \ldots, D_M\}$ of quadruples (a, b, c, d) of positive integers. We wish to study assignments A of these quadruples to all the lattice points of the first infinite quadrant of the Cartesian plane, or mappings A of the set N^2 of ordered pairs of nonnegative integers into the set P, such that:

4.1
$$a(Axy') = c(Axy),$$
$$b(Ax'y) = d(Axy),$$

where x' is short for $x + 1$ and $a(D)$, $b(D)$, $c(D)$, $d(D)$ are respectively the first, second, third, fourth members of the quadruple D. The first question is whether there is a general procedure by which, given any finite set P, we can decide whether there exists an assignment satisfying the condition 4.1.

It is natural to use the ordinary Cartesian coordinates and identify each unit square with the point at its lower left hand corner. Then we can speak of the origin $(0, 0)$, the main diagonal $x = y$, etc.

The following problems are all known to be unsolvable:

4.2   *The (unrestricted) domino problem.* To find an algorithm to decide, for any given (finite) set of domino types, whether it is solvable.

4.3   *The origin-constrained domino problem.* To decide, for any given set P of domino types and a member C thereof whether P has a solution with the origin occupied by a domino of type C.

4.4   *The diagonal- (row-, column-) constrained domino problem.* To decide, for any given set P of domino types and a subset Q thereof, whether P has a solution with the main diagonal (the first row, the first column) occupied by dominoes of types in Q.

We shall give here only a proof of the simplest case 4.3, which was obtained in 1960 soon after the problems had been formulated.

In order to prove the unsolvability of the origin-constrained domino problem, we shall reduce to it the following familiar unsolvable halting problem of Turing machines.

(HB)   To decide, given any Turing machine, whether it eventually halts if the initial tape is blank. (see 3.6.3).

To keep our ideas fixed, we shall assume one special formulation of Turing machines, although it will be clear that similar considerations are applicable to other formulations. We use a one-way infinite tape, take q_1 as the initial state, the leftmost square of the tape as the initially scanned square, two tape symbols S_0 (blank) and S_1 (marked), the basic acts R (shift the reading head right one square), L (shift the reading head left one square), S_1 (print S_1), S_0 (print S_0). Each machine has a finite number of states $q_1, \ldots,$ q_n, and, at each moment, the present state and the content of the scanned square together determine the acts (one print act and one shift act) to be taken, as well as the state at the next moment.

An example which will be used for illustration is:
Machine X.

$$q_1 S_0 S_1 R q_2 \qquad\qquad q_1 S_1 S_1 R q_1$$
$$q_2 S_0 S_0 R q_3 \qquad\qquad q_2 S_1 S_1 L q_3$$
$$q_3 S_0 S_1 L q_4 \qquad\qquad q_3 S_1 S_0 L q_4$$
$$q_4 S_0 S_0 L q_1$$

We shall give a general method by which, given any Turing machine X, we can find a corresponding domino set P_X containing a distinguished type D such that X halts on an initially blank tape if and only if P has no solution with a domino of type D at the origin. Essentially, we choose P_X so that in a solution of P, for every y, the yth row contains the whole situation of X (tape, state, and scanned square) at time y. As a result X eventually halts if and only if P_X has no solution.

Thus, for the example X above, we choose P_X as follows.

4.5 P_X consists of the following domino types:

4.5.1   Two domino types for each tape symbol:

$$[S_0], [LS_0], [S_1], [LS_1].$$

4.5.2   One domino type for each permissible kind of scanned square (state and symbol): $[q_i S_j]$, $i = 1, 2, 3, 4$; $j = 0, 1$; $(i, j) \neq (4, 1)$.

4.5.3 One domino type for the next scanned square (symbol and next state) after a left shift: $[Lq_iS_j]$, $i = 1, 3, 4$; $j = 0, 1$.

4.5.4 One type for the next scanned square after a right shift: $[Rq_iS_j]$, $i = 1, 2, 3$; $j = 0, 1$.

4.5.5 Four domino types for the initial row and column: $[D]$ for the origin, $[B]$ for the beginning of the tape, $[\uparrow]$ for initial row, $[\rightarrow]$ for the initial column.

If we use the x-coordinate to represent tape positions and the y-coordinate to represent time, a simulation of the particular machine X should be given by a partial solution of P_X as in Figure 1. Machine X halts at $y = 8$, because the reading head ends up scanning S_1 in state q_4, and it is understood that the machine halts if either there is left shift while scanning the beginning of the tape, or the reading head scans S_j in state q_i but there is no entry in the machine table beginning with q_iS_j.

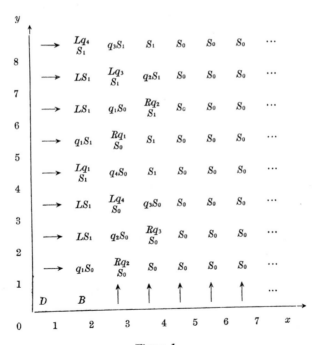

Figure 1

There remains the problem of specifying the four numbers or colors of each domino type in P_x to exclude undesired solutions.

Each solution is an assignment or mapping A of a domino type to a member of N^2. It is natural to write $Axy = D_i$ briefly as $D_i xy$. We can now state the conditions needed for the simulation.

4.6   *The conditions on P_x.*

4.6.1   *The origin constraints. (Ex) $[D]xx$.*

In other words, the type $[D]$ must occur somewhere. This position is treated as the origin $(0, 0)$. By choosing the colors on D suitably, we can make it impossible for any domino to occur to its left or below it.

4.6.2   The initial row and the initial column are the boundary:

(1) $[D]xy \supset [B]x'y$.

(2) $([B] \vee [\uparrow])xy \supset [\uparrow]x'y$.

(3) $([D] \vee [\rightarrow])yx \supset [\rightarrow]yx'$.

4.6.3   The next row above the initial row simulates the initial configuration:

(1) $[B]yx \supset [q_1 S_0]yx'$.

(2) $[\uparrow]yx \supset ([Rq_2 S_0] \vee [S_0])yx'$.

4.6.4   The left or right neighbor of the scanned square at time y is in part determined by a left or right shift and embodies information for the scanned square at time y'.

It is convenient to write briefly $[Lq_i]$ for $[Lq_i S_0] \vee [Lq_i S_1]$, $[Rq_i]$ for $[Rq_i S_0] \vee [Rq_i S_1]$.

(1) $[q_i S_j]x'y \supset [Lq_k]xy$; $(i, j, k)=(2, 1, 3), (3, 0, 4), (3, 1, 4),$ $(4, 0, 1)$.

(2) $[q_i S_j]xy \supset [Rq_k]x'y$; $(i, j, k)=(1, 0, 2), (1, 1, 1), (2, 0, 3)$.

4.6.5   The state and scanned square at time y' are determined by $[Lq_i]$ or $[Rq_i]$ at time y.

(1) $[Lq_i S_j]yx \supset [q_i S_j]yx'$; $i = 1, 3, 4,$ $j = 0, 1$.

(2) $[Rq_i S_j]yx \supset [q_i S_j]yx'$; $i = 1, 2, 3,$ $j = 0, 1$.

4.6.6   The tape symbol at time y' and position x is determined by the tape symbol at (x, y).

(1) $[S_i]yx \supset ([S_i] \vee [Rq_1 S_i] \vee [Rq_2 S_i] \vee [Rq_3 S_i])yx'$;
$i = 0, 1$.
$[LS_i]yx \supset ([LS_i] \vee [Lq_1 S_i] \vee [Lq_3 S_i] \vee [Lq_4 S_i])yx'$;
$i = 0, 1$.

(2) $[q_2S_0]yx \supset [Lq_4S_0]yx'$,

  $[q_3S_1]yx \supset [S_0]yx'$,

  $[q_4S_0]yx \supset ([Rq_1S_0] \vee [Rq_2S_0])yx'$,

(3) $[q_1S_0]yx \supset ([LS_1] \vee [Lq_3S_1])yx'$,

  $[q_1S_1]yx \supset [LS_1]yx'$,

  $[q_2S_1]yx \supset [S_1]yx'$,

  $[q_3S_0]yx \supset [S_1]yx'$.

4.6.7 In each row, we have to distinguish $[S_i]$ and $[LS_i]$.

(1) $([Rq_1] \vee [Rq_2] \vee [Rq_3] \vee [q_2S_1] \vee [q_3S_0] \vee [q_3S_1]$
   $\vee [q_4S_0] \vee [S_0] \vee [S_1])xy \supset ([S_0] \vee [S_1])x'y$.

(2) $([Lq_1] \vee [Lq_3] \vee [Lq_4] \vee [q_1S_0] \vee [q_1S_1] \vee [q_2S_0]$
   $\vee [LS_0] \vee [LS_1])x'y \supset ([LS_0] \vee [LS_1] \vee [\rightarrow])xy$.

4.6.8 *The halting conditions.*

(1) $\neg[q_4S_1]xy$.

We can either exclude the type $[q_4S_1]$ or choose its colors so that no domino can occur above it.

(2) $[\rightarrow]xy \supset \neg([Lq_1] \vee [Lq_3] \vee [Lq_4])xy$.

This could be deleted if we had included the condition that no two types can be assigned to the same place.

These conditions determine the colors on the domino types in the following manner.

4.7 Colors on the domino types in P_x are given in Figure 2.

By 4.6.5, the top of Lq_iS_j or Rq_iS_j is the same as the bottom of q_iS_j. By (1) in 4.6.3, the top of B is the same as the bottom of q_1S_0.

By (1) in 4.6.6, the bottom of the first three rows and the top of the first row are filled. By (2) in 4.6.3, the top of $[\uparrow]$ is determined. By (3) in 4.6.6, the top of the third and the fourth rows is determined.

By (1) and (2) in 4.6.7, the left of the first two rows, the right of the first and the third rows, the L's and R's in the third and the fourth rows, and the right of $[\rightarrow]$ are determined.

By 4.6.4, the remaining edges in the middle four rows are determined.

Finally, by 4.6.2, the edges of the last row are determined.

The conditions 4.6 do not exclude the possibility that several domino types occur at the same place, although the condition 4.6.1

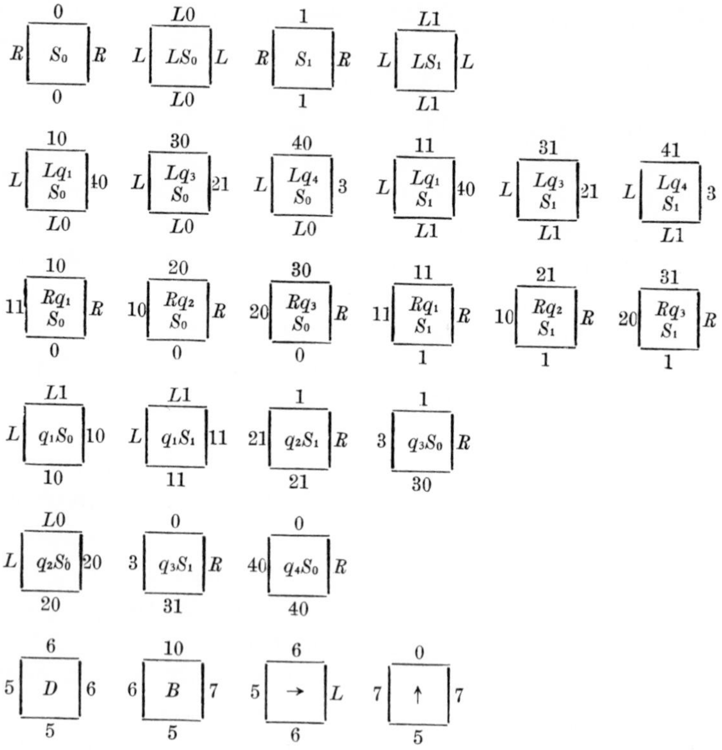

Figure 2 (dominoes):

Row 1:
- 0 / R | S_0 | R / 0
- L0 / L | LS_0 | L / L0
- 1 / R | S_1 | R / 1
- L1 / L | LS_1 | L / L1

Row 2:
- 10 / L | Lq_1S_0 | 40 / L0
- 30 / L | Lq_3S_0 | 21 / L0
- 40 / L | Lq_4S_0 | 3 / L0
- 11 / L | Lq_1S_1 | 40 / L1
- 31 / L | Lq_3S_1 | 21 / L1
- 41 / L | Lq_4S_1 | 3 / L1

Row 3:
- 10 / 11 | Rq_1S_0 | R / 0
- 20 / 10 | Rq_2S_0 | R / 0
- 30 / 20 | Rq_3S_0 | R / 0
- 11 / 11 | Rq_1S_1 | R / 1
- 21 / 10 | Rq_2S_1 | R / 1
- 31 / 20 | Rq_3S_1 | R / 1

Row 4:
- L1 / L | q_1S_0 | 10 / 10
- L1 / L | q_1S_1 | 11 / 11
- 1 / 21 | q_2S_1 | R / 21
- 1 / 3 | q_3S_0 | R / 30

Row 5:
- L0 / L | q_2S_0 | 20 / 20
- 0 / 3 | q_3S_1 | R / 31
- 0 / 40 | q_4S_0 | R / 40

Row 6:
- 6 / 5 | D | 6 / 5
- 10 / 6 | B | 7 / 5
- 6 / 5 | → | L / 6
- 0 / 7 | ↑ | 7 / 5

Figure 2

does yield the requirement that at least one domino type occurs at each place. To assure uniqueness, we may either add an explicit condition to such an effect or use $\equiv$ instead of $\supset$. If we replace $\supset$ by $\equiv$, we have to put together occurrences of the same basis formula, such as $[q_1S_0]yx'$ in (1) of 4.6.3 and (1) of 4.6.5 $([Lq_1S_0] \lor [Rq_1S_0] \lor [B])yx \equiv [q_1S_0]yx'$. Such a treatment is perhaps less easy to follow.

Since the example is sufficient to illustrate the general situation, we have proved:

Theorem 1.

The origin-constrained domino problem is unsolvable; in fact, the halting problem HB is reducible to it.

The original application of these results is to the decision problem of first order logic (see Chapter 5). We indicate how to correlate tiling problems with logical formulas.

Given a set $P = \{D_1, \ldots, D_M\}$, we may divide the set into four sets P_1, P_2, P_3, P_4 of subsets such that two dominoes belong to the same subset in P_1 if and only if their first members (bottom edges) are the same (of the same color), and so on. Then we can match up subsets of P_1 with those of P_3, subsets of P_2 with those of P_4 so that, e.g., a subset $\{D_e, \ldots, D_f\}$ in P_4 is matched up with a subset $\{D_g, \ldots, D_h\}$ in P_2 if the fourth member of D_e is the same as the second member of D_g.

4.8   A mapping of N^2 into P gives a solution if and only if:

4.8.1   Every pair (x, y) in N^2 gets a unique quadruple from P; or, writing $D_i xy$ as short for $Axy = D_i$:

   (1) $D_1 xy \vee \cdots \vee D_M xy$.

   (2) $[D_1 xy \supset (\neg D_2 \wedge \cdots \wedge \neg D_M)xy] \wedge \cdots \wedge [D_M xy \supset (\neg D_1 \wedge \cdots \wedge \neg D_{M-1})xy]$.   Briefly, 4.8.1, i.e., the conjunction of (1) and (2), is also written as: $\nabla(D_1 xy, \cdots, D_M xy)$.

4.8.2 $b(Ax'y) = d(Axy)$.   Or, using truth functions only:

$$[(D_e \vee \cdots \vee D_f)xy \equiv (D_g \vee \cdots \vee D_h)x'y]$$
$$\wedge \cdots \wedge [(D_p \vee \cdots \vee D_q)xy \equiv (D_s \vee \cdots \vee D_t)x'y].$$

4.8.3 $a(Ayx') = c(Ayx)$, with a similar truth functional expression of the form $V(D_1 yx, \ldots, D_M yx; D_1 yx', \cdots, D_M yx')$.

More exactly, we have assumed that in the set P, for every domino D_i, there are D_j, D_k, such that $b(D_j) = d(D_i)$, $a(D_k) = c(D_i)$, because we can effectively eliminate domino types D_i for which the above condition does not hold. The fact that we are concerned with the first quadrant rather than the whole plane presents a slight complication in so far as a type D_i for which there is no D_j, $a(D_i) = c(D_j)$, or no D_k, $b(D_i) = d(D_k)$ may occur in the first row or the first column but not elsewhere. It is, however, clear that all members of each $P_k(k = 1, 2, 3, 4)$ are mutually exclusive and jointly exhaust P. Moreover, it is true in every case that in 4.8.2 and 4.8.3 for each D_i, $D_i xy$ or $D_i x'y$ or $D_i yx$ or $D_i yx'$ appears in at most one disjunct because, e.g., if $c(D_i) = a(D_j) = a(D_k)$ and $c(D_p) = a(D_j)$, we must also have $c(D_p) = a(D_k)$.

In condition 4.8, we can, by the help of (2) in 4.8.1, replace $\equiv$ by $\supset$ in 4.8.2 and 4.8.3. Another possibility is to replace $\equiv$ by $\wedge$, and $\wedge$ by $\vee$ in 4.8.2 and 4.8.3. Then we can delete (1) in 4.8.1

as an independent condition.

From the definition of solvability, it follows immediately that to every set P of domino types, there is a corresponding formula of the form:

(1) $U(G_1xy, \cdots, G_kxy; G_1x'y, \cdots, G_kx'y) \wedge V(G_1yx, \cdots, G_kyx; G_1yx', \ldots, G_kyx')$, or briefly, and using a suitable form of the Löwenheim-Skolem Theorem[1]. (1) $U(xy, x'y) \wedge V(yx, yx')$, or $\forall x \exists u \forall y [U(xy, uy) \wedge V(yx, yu)]$ where U and V are truth-functional combinations of the components, such that P is solvable if and only if (1) has a model, i.e., there is an interpretation in the domain N of nonnegative integers of $G_1, \ldots, G_k$ which makes (1) true.

Note incidentally, we can generally use fewer dyadic predicates than dominoes. Given M domino types, let $K = \mu n(2^n \geqslant M)$. We can then use K dyadic predicates to represent the M domino types, since for any x and y, each of $G_1xy, \ldots, G_kxy$ can be true or false and we can identify each D arbitrarily with one of these 2^k distributions of truth values, e.g., replace D_ixy by $G_exy \wedge \cdots \wedge G_fxy \wedge \neg G_gxy \wedge \cdots \wedge \neg G_hxy$. Using any such representation, we can restate 4.8, e.g., by:

4.9 The set $\{D_1, \ldots, D_M\}$ is solvable if and only if we can assign values to $G_1, \ldots, G_k$ over all (x, y) in N^2, such that:

4.9.1 $[(D_exy \vee \cdots \vee D_fxy) \wedge (D_gx'y \vee \cdots \vee D_hx'y)] \vee \cdots$
$\vee [(D_pxy \vee \cdots \vee D_qxy) \wedge (D_sx'y \vee \cdots \vee D_tx'y)]$.

4.9.2 Similarly for (y, x) and (y, x').

The uniqueness of (2) in 4.8.1 is now dispensable since the truth distributions of $G_1xy, \ldots, G_kxy$ are automatically mutually exclusive. Moreover, as just noted, the existence of condition (1) in 4.8.1 also follows from 4.9.1 (or 4.9.2) since no matter which disjunct of 4.9.1 (or 4.9.2) is true, some D_i must be true of (x, y).

We shall now prove that, conversely, given a formula F of the form (1), we can find a corresponding domino set P_F such that P_F is solvable if and only if F has a model in N.

We assume F contains k dyadic predicates $G_1, \ldots, G_k$ and U, V are in the fully developed disjunctive normal form so that U (or V)

---

1) See Chapter 5, section 5 2 for more discussions and references (in particular, references under footnote 2) on page 258.

is a disjunction of conjunctions each of which is of the form

(2) $\quad \pm G_1 xy \wedge \cdots \wedge \pm G_k xy \wedge \pm G_1 x'y \wedge \cdots \wedge \pm G_k x'y$ or

(2*) $\quad \pm G_1 yx \wedge \cdots \wedge \pm G_k yx \wedge \pm G_1 yx' \wedge \cdots \wedge \pm G_k yx'$.

If pq is ambiguously xy, yx, $x'y$, or yx', we separate out the sign pattern of each occurring K-termed conjunction $\pm G_1 pq \wedge \cdots \wedge \pm G_K pq$ by writing it as $C_i pq$, one number i for each pattern. Each of the patterns C_i is taken as a color and we define the set P_F of dominoes as the set of all quadruples $D_K = (C_i, C_i, C_j, C_k)$ such that $[(C_i yx \wedge C_j yx') \supset V(yx, yx')]$ and $[(C_i xy \wedge C_k x'y) \supset U(xy, x'y)]$ are truth-functional tautologies.

In this way, each sign pattern C_i gives rise to $m \times n$ domino types if it can be followed on the top by m sign patterns, and on the right by n sign patterns. Had we taken each C_i as a single domino type, we would not be able to exclude in general the situation that, e.g., C_1 can be followed by C_2 and C_3 on the right, but C_4 can only be followed by C_2 on the right, violating the requirement on colors.

If F has a model, then the conjunction of all instances of $U(xy, x'y) \wedge V(yx, yx')$ for all (x, y) in N^2 is true for some selection of a conjunction (2) and a conjunction (2*), for each pair (x, y). This then yields a solution of P_F.

Conversely, if P_F has a solution, we obtain from the solution two conjunctions (2) and (2*) for each pair (x, y). All these selections yield together a model for the given formula F.

Hence, we get:

Theorem 2.

> Given a domino set P we can find a formula F_P of the form (1) such that P has a solution if and only if F_P has a model; conversely, given a formula F of the form (1), we can find a domino set P_F such that F has a model if and only if P_F has a solution. Hence, the unrestricted domino problem is undecidable if and only if the decision problem of the class of all formulas of the form (1) is unsolvable.

5 Tag systems and lag systems

A combinatorial system in the most general sense would be any finite set of rules each of which effectively produces a finite set of conclusions from a finite set of premises. The most intensively studied case is the one in which each rule has a single premise and a single conclusion. Such a system is called monogenic if the rules are such that for any string at most one rule is applicable.

From this broad class of monogenic systems, Post[1] chooses to consider the tag systems. A tag system is determined by a finite set of rules:

$$T_i: \ s_i \longrightarrow E_i, \qquad i = 1, \cdots, \varrho,$$

such that if the first symbol of a string is s_i, then the first β symbols are removed and the string E_i is appended at the end. Since the system is monogenic, $s_i \neq s_j$ when $i \neq j$. If the alphabet contains σ symbols, then $\varrho \leq \sigma$.

Another natural class is, for want of a better name, the lag systems. A lag system is a set of $\leq \sigma^\beta$ rules:

$$L_i: \ s_{i_1} \cdots s_{i_\beta} \longrightarrow E_i,$$

such that if the first β symbols of a string are $s_{i_1} \cdots s_{i_\beta}$, the first symbol, viz., s_{i_1}, is deleted and E_i is appended at the end of the string. In either kind of system, E_i is permitted to be the null string. Let ε_i be the length of E_i. ε be the maximum and ε^- be the minimum of ε_i. Thus each system is associated with three constants, β, ε, σ, which, in general, appear to be of decreasing importance in that order. Clearly, when $\beta = 1$, tag systems and lag systems coincide. In general, lag systems are less wasteful since no symbol in a string is overlooked.

We shall establish, in a sense to be specified, that certain tag systems and lag systems are "undecidable", and all "simpler" ones are decidable. The undecidable tag system is a slight improvement over the one constructed by John Cocke and Minsky[2] with $\beta = 2$, $\varepsilon = 4$, to one with $\beta = 2$, $\varepsilon = 3$.

5.1   All monotone systems are decidable.

With regard to each (tag or lag) system, there is a halting

---

1)   E. Post, *Am. journal math.*, vol. 65 (1943), pp. 197—268.

2)   See M. Minsky, *Annals of math.*, vol. 74 (1961), pp. 437—454 and p. 270 **of his book listed under footnote on page 242.**

problem and a derivability problem. A system halts on a given string S if S ever leads to a string Q whose length $|Q| < \beta$, or to which no rule in the given system is applicable. The halting problem is to give a general method to decide, for each given string from the alphabet of the system, whether the system halts on the string. The derivability problem is to decide, for any two given strings, whether the rules will lead us from the first to the second.

If we follow Post[1] in requiring $\rho = \sigma$ for tag systems and analogously $\rho = \sigma^\beta$ for lag systems, then, since there are only finitely many strings of length less than β, a positive solution of the derivability problem yields one for the halting problem, and a negative solution of the latter yields one for the former. In permitting $\rho < \sigma$ or σ^β, we do not have such a simple connection between the two problems. We shall confine ourselves to the more restricted tag and lag systems.

It is quite evident that the decision problems become complex only when some rules expand a string while others contract it. This remark can be stated and justified more exactly in two theorems.

Theorem 1. For any given tag system T, if $\beta \geqq \varepsilon$ or $\beta \leqq \varepsilon^-$, then the derivability (and hence also the halting) problem for T is decidable.

Theorem 2. For any given lag system L, if $\varepsilon \leqq 1$ or $\varepsilon^- \geqq 1$, then the derivability (and hence also the halting) problem for L is decidable. This includes all lag systems in which no E_i is the null string.

The proofs are similar. We give only a proof of Theorem 1.

Let $\beta \leqq \varepsilon^-$ and S be a given string. If $\beta < \varepsilon^-$, then, for each $n \geqq |S|$, there is at most a single consequence of S by T that is of length n, and there is no consequence shorter than S. Suppose, for some $i, \beta = \varepsilon_i$. Each time such a "stable" rule is applied, the length of the sequence does not change. Let $|S| = a$, $|Q| = b$, $b - a = c$. If c is negative, then of course Q is not a consequence of S. Otherwise, we write out the successive consequences of S one by one, $S = S_0$, S_1, S_2, etc., until we obtain either a repetition or a string longer than Q. Clearly $|S_i| \leqq |S_{i+1}|$, for all i. Moreover, for each fixed length k, there can be at most σ^k strings of length k. Hence, the process must always terminate. If $S_p = S_q$, q > p, and Q is not among $S_0, \ldots, S_{q-1}$ then Q is not a consequence of S because $S_p, \ldots, S_{q-1}$ will repeat and all consequences are contained in $S_0, \ldots, S_{q-1}$. If now $|S_t > Q$ and Q is not among $S_0, \ldots, S_t$, then again Q is not a consequence of S.

Suppose $\beta \geqq \varepsilon$ and S is a given string. The argument is similar except that in this case we can list, once and for all, all consequences of S. Thus, either we get $S_q = S_p$, $p < q$, then $S_0, \ldots, S_{q-1}$ are all the consequences; or else, we get S_t which is the null sequence, and then $S_0, \ldots, S_{t-1}$ are all the consequences.

5.2   Every system with $\beta = 1$ is decidable.

Since there is no distinction between tag systems and lag systems for $\beta = 1$, we shall speak only in terms of tag systems. We assume therefore a tag system T with rules:

$$R_i: \ s_i \longrightarrow E_i, \qquad i = 1, \cdots, \sigma.$$

Since $\beta = 1$, the only contraction rules are those which produce the null string. By Theorem 1, we only have to consider the case when there are contraction rules.

5.2.1   *Definition of ranks.* If E_i is null, the rule R_i and the symbol s_i are of rank 1. If E_i is not null, but every symbol in E_i is of finite rank, then the rank of s_i and R_i is $n + 1$. n being the maximum of the ranks of the symbols in E_i. If a rule R_i does not get a rank in the above manner, then R_i and s_i are said to have an infinite rank. Clearly:

5.2.2   If every rule of T has a finite rank, then every string S has only finitely many consequences. If some rules have the infinite rank but S contains only symbols with finite ranks, S again has only finitely many consequences.

It may be noted that the halting problem is easy to decide. Thus, given a tag system T and a string S, let T^* be obtained from T by deleting first all rules of finite ranks, and then all symbols of finite ranks from the remaining rules. Similarly, let S^* be obtained from S by deleting all symbols of finite ranks. T halts on S if and only if S^* is null. Thus, if S^* is not null, S can keep on producing consequences by T^* and therefore by T. On the other hand, if S^* is null, then there can be only finitely many consequences.

To decide the derivability problem, we introduce more definitions.

5.2.3 Let R_i be of infinite rank. Consider the σ consequences $A_1, \ldots, A_\sigma$ of s_i, with all symbols of finite ranks deleted. If some A_j contains s_i as a proper part, then R_i and s_i are of finite degree. If at least one A_j contains at least one symbol of finite degree, then R_i and s_i are of finite degree. Otherwise, R_i is a circular rule and s_i is a circular symbol.

5.2.4 Given a string S and a tag system T, we have made a round if we have operated on every symbol in the string. The next round takes the result of the previous round as the given string. If T contains η circular rules, a circular symbol s_i is periodic if beginning with s_i as the initial string, we arrive, after η or less rounds, at a string in which s_i is again the only symbol of infinite rank.

To clarify these definitions, we observe the following. If T contains no circular rules, then, beginning with a string S containing some symbols of infinite rank (other strings being trivial by 5.2.2), we must get a string with more symbols with infinite ranks, after at most σ rounds. Hence, if $|Q| = m$, either Q occurs among the first σm rounds of consequences of S, or Q is not a consequence of S. Moreover, even when T contains circular rules, if S contains symbols of finite degree, the considerations still hold. Hence, we have to consider only strings containing no symbols of finite degree in systems containing circular rules.

Given one such system T, there must be at least one periodic symbol s_i. Thus, beginning with any circular s_j, we can never encounter a symbol of finite degree because otherwise, s_j itself would be of finite degree. Hence, at each stage there is exactly one circular symbol. Since there are only finitely many (certainly $\leqq \sigma$) circular symbols, at least one of them must be a periodic symbol.

5.2.5 Beginning with a periodic symbol a_i as the initial string, we can always find in less than $2\sigma^2$ rounds two consequences S_p, S_q, such that $p \neq q$ but $S_p = S_q$.

The only complication is with symbols of finite ranks since at each stage there is exactly one symbol of infinite rank. If now, beginning from a_i, we come, after enough (say $t \leq \sigma$) rounds, for the first time to a string A_i containing a_i again, say $x_1 \ldots x_u a_i y_1 \ldots y_v$, then $x_1, \ldots, x_u, y_1, \ldots, y_v$ must all be of finite ranks. Call t the period of a_i. From A_i on, after each t rounds, we get another sequence which contains A_i as a (proper or improper) part, since a_i always produces A_i after t rounds. The maximum finite rank of the rules of T is $\leq \sigma t$, and if B_i is obtained from A_i after $\sigma t (\leq \sigma^2)$ rounds, then, B_i is again obtained after σt more rounds. This is so because, although B_i may be $C_i A_i D_i$, C_i and D_i can have no more effect after σt rounds and the result is entirely determined by A_i.

We are now ready to settle the principal case. Thus, T contains circular rules, and each symbol in S is either of finite rank or circular. We wish to decide whether an arbitrary string Q is a consequence of S. Note first that after at most σ rounds, we eliminate all the circular symbols which are not also periodic. Hence, if t is the least common multiple of the periods $t_1, \ldots, t_j$ of all the periodic elements, then, by 5.2.5 after $(2\sigma + 1)t$ rounds, we get a repetition and therefore the set of all consequences of S.

Hence we have proved:

Theorem 3. The derivability problem for each tag system or lag system with $\beta = 1$ is decidable.

5.3   *Undecidable lag systems.*

We shall give a lag system with $\beta = \varepsilon = 2$ whose halting problem is undecidable, by using SS machines introduced in Shepherdson-Sturgis[1]. They have shown that every Turing machine (in particular, a universal one) can be simulated by an SS machine on the alphabet $\{0, 1\}$. We shall give a procedure of simulating these SS machines.

An SS machine is a finite sequence of instructions each of which is of the following two types.

P_0, P_1: print 0 (or 1) at the right end of the string S and go to the next instruction.

$SD(k)$: scan and delete the leftmost symbol of S; if it is 0, go to the next instruction, otherwise go to instruction k; if S is null,

---

1)   See *Journal ACM*, vol. 10 (1963), pp. 217—255.

halt.

Let $q_1, \ldots, q_n$ be the instructions of an SS machine working on strings from $\{0, 1\}$. If the initial string is $x_1 \ldots x_p$, we shall represent it by $b_i x_1 \ldots x_p e_i$. At each stage, if the state or instruction is q_i, the working string is of the form $b_i x_1 \ldots x_p e_i$. Consider now any instruction q_i. For brevity, let $j = i + 1$.

For the n instructions, we have altogether $2n + 4$ symbols $b_1,$ $\ldots, b_{n+1}, e_1, \ldots, e_{n+1}, 0, 1$.

Case 1. q_i is P_0. We wish to get from $b_i x_1 \ldots x_p e_i$ to $b_j x_1 \ldots x_p 0 e_j$. The rules are:

$$(1) \qquad x_t x_{t+1} \longrightarrow x_{t+1} \qquad (x_t, x_{t+1} = 0 \text{ or } 1),$$

$$(2) \qquad b_i x \longrightarrow b_j x \qquad (x = 0 \text{ or } 1),$$

$$(3) \qquad e_i b_j \longrightarrow e_j$$

$$(4) \qquad x e_i \longrightarrow 0 \qquad (x = 0 \text{ or } 1).$$

Hence, by (2), $\qquad b_i x_1 \cdots x_p e_i \longrightarrow x_1 \cdots x_p e_i b_j x_1,$

by (1), $\qquad \longrightarrow x_p e_i b_j x_1 \cdots x_p,$

by (4), $\qquad \longrightarrow e_i b_j x_1 \cdots x_p 0,$

by (3), $\qquad \longrightarrow b_j x_1 \cdots x_p 0 e_j.$

In order to cover the case $p = 0$, we add the rule:

$$(5) \qquad b_i e_i \longrightarrow b_j 0.$$

Hence, by (5), $\qquad b_i e_i \longrightarrow e_i b_j 0,$

by (3), $\qquad \longrightarrow b_j 0 e_j.$

Case 2. q_i is P_1. The rule (1) above plus:

$$(6) \qquad b_i x \longrightarrow b_j x,$$

$$(7) \qquad e_i b_j \longrightarrow e_j,$$

$$(8) \qquad x e_i \longrightarrow 1,$$

$$(9) \qquad b_i e_i \longrightarrow b_j 1.$$

Case 3. q_i is $SD(k)$. We wish to get from $b_i x_1 \ldots x_p e_i$ to $b_j x_2 \ldots x_p e_j$ if $x_1 = 0$. to $b_k x_2 \ldots x_p e_k$ if $x_1 = 1$. The rules are, besides (1) above:

$$(10) \qquad b_i 0 \longrightarrow b_j,$$

$$(11) \qquad b_i 1 \longrightarrow b_k,$$

(12)
$$xe_i \longrightarrow \ ,$$

(13)
$$e_i b_j \longrightarrow e_j,$$

(14)
$$e_i b_k \longrightarrow e_k,$$

(14*)
$$b_i e_i \longrightarrow \text{(redundant)}.$$

The last rule covers the trivial case when the string is null. The SS machines halt then. For the nontrivial case, we have:

Case 3.1. $x_1 = 0$

By (10), $\quad b_i 0 x_2 \cdots x_p e_i \longrightarrow 0 x_2 \cdots x_p e_i b_j,$

by (1), $\quad\qquad\qquad\qquad \longrightarrow x_p e_i b_j x_2 \cdots x_p,$

by (12), $\quad\qquad\qquad\qquad \longrightarrow e_i b_j x_2 \cdots x_p,$

by (13), $\quad\qquad\qquad\qquad \longrightarrow b_j x_2 \cdots x_p e_j.$

Case 3.2. $x_1 = 1$

By (11), $\quad b_i 1 x_2 \cdots x_p e_i \longrightarrow 1 x_2 \cdots x_p e_i b_k,$

by (1), $\quad\qquad\qquad\qquad \longrightarrow x_p e_i b_k x_2 \cdots x_p,$

by (12), $\quad\qquad\qquad\qquad \longrightarrow e_i b_k x_2 \cdots x_p,$

by (14). $\quad\qquad\qquad\qquad \longrightarrow b_k x_2 \cdots x_p e_k.$

There are two remaining loose ends to tidy up. If $i = n$, then $j = i + 1 = n + 1$, and in $SD(k)$, we may have $k > n$. In these cases, the SS machine is to halt. Clearly when $k > n$ in $SD(k)$, we can always put $k = n + 1$. Hence, the necessary rule needed is to halt when we encounter $b_{n+1} x_1 \ldots x_p e_{n+1}$. This is taken care of by the following rules:

(15)
$$b_{n+1} x \longrightarrow b_{n+1},$$

(16)
$$x e_{n+1} \longrightarrow e_{n+1},$$

(17)
$$e_{n+1} b_{n+1} \longrightarrow \ ,$$

(18)
$$b_{n+1} e_{n+1} \longrightarrow \ .$$

Hence, by (15), $b_{n+1} x_1 \cdots x_p e_{n+1} \longrightarrow x_1 \cdots x_p e_{n+1} b_{n+1},$

$\qquad$ by (1), $\qquad\qquad\qquad\qquad \longrightarrow x_p e_{n+1} b_{n+1} x_2 \cdots x_p,$

$\qquad$ by (16) and (17), $\qquad\quad \longrightarrow b_{n+1} x_2 \cdots x_p e_{n+1},$

$\qquad$ repeating, $\qquad\qquad\qquad \longrightarrow b_{n+1} e_{n+1},$

$\qquad$ by (18), $\qquad\qquad\qquad\quad \longrightarrow e_{n+1}.$

Hence, the lag system halts by definition.

The other loose end is that we have not used all the $(2n + 4)^2$ pairs of symbols in the rules, e.g., $e_i e_k$. This is a simple matter since such combinations of letters do not occur. We add simply:

(19) $cd \rightarrow$, if cd does not begin any of the rules (1) to (18).

Therefore, we have established:

Theorem 4. There is a lag system with $\beta = \varepsilon = 2$, whose halting problem is unsolvable.

By Theorem 3, lag systems with $\beta = 1$ are always decidable. By Theorem 2, when $\beta = 2$, a lag system is decidable if either no $\varepsilon_i \geqq 2$ or no $\varepsilon_i = 0$. Hence, if we disregard the less decisive factor, the number σ of symbols in the alphabet, the above result is the best possible.

5.4 *Undecidable tag systems.*

We modify somewhat an earlier proof[1] to get the following theorem.

Theorem 5. There is a tag system with $\beta = 2$, $\varepsilon = 3$, $\varepsilon^- = 1$, whose halting problem is unsolvable.

By Theorem 3 and Theorem 1, we can say this is the best possible result. It is perhaps of interest to compare this with the undecidable lag system: while we need here $\varepsilon = 3$ rather than 2, we do not need rules with the null string as consequences.

In order to prove the theorem, we use a program formulation of Turing machines[2]. Each machine is represented by a finite sequence of instructions (or states) of the following five types: M (mark), E (erase), L (shift left), R (shift right), Ck (transfer to instruction k if the square under scan is marked, otherwise go to next instruction). Compared with the earlier proof, the present version separates the different actions (shift, write, transfer) so that the argument is perhaps easier to follow.

Assume now we are given a universal Turing machine and its program, with the two symbols 0, 1 only. The halting problem is unsolvable with initial inputs confined to include only finitely many 1's. At instruction or state i, the whole configuration is represented by $(x_i x_i)^m s_i s_i (y_i y_i)^n$, where the minimum portion including all 1's

---

1) See footnote on page 258.
2) See paper listed under footnote on page 242.

and the symbol s_i under scan is $\cdots a_2 a_1 a_0 s b_0 b_1 b_2 \cdots$, and $m = \Sigma a_t 2^t$, $n = \Sigma b_t 2^t$. The exponention on $x_i x_i$ and $y_i y_i$ means repetition so that, e.g., $(x_i x_i)^0$ is the empty string.

For each state i, a corresponding alphabet is introduced with tag rules which produce $(x_j x_j)^{m'} s_j s_j (y_j y_j)^{n'}$ from $(x_i x_i)^m s_i s_i (y_i y_i)^n$, j being the next state. The simulation for all instructions of a kind is uniform. Except for Ck, j is always $i + 1$, although we could as well take any fixed j in each case, and Ck is the only kind of instruction that introduces branching.

(a)   Mark: $i.M$. This leads from $(x_i x_i)^m s_i s_i (y_i y_i)^n$ to

$$(x_{i+1} x_{i+1})^m 1_{i+1} 1_{i+1} (y_{i+1} y_{i+1})^n.$$

$$x_i \longrightarrow x_{i+1} x_{i+1},$$
$$s_i \longrightarrow 1_{i+1} 1_{i+1},$$
$$y_i \longrightarrow y_{i+1} y_{i+1}.$$

(b)   Erase: $i.E$. Similar, replace second rule by: $s_i \to 0_{i+1} 0_{i+1}$.

(c)   Conditional transfer: $i.Ck$. Given $(x_i x_i)^m s_i s_i (y_i y_i)^n$, this goes to $(x_{i+1} x_{i+1})^m s_{i+1} s_{i+1} (y_{i+1} y_{i+1})^n$ if s_i is 0_i, and to $(x_k x_k)^m s_k s_k (y_k y_k)^n$ otherwise.

This case gives a simple illustration of the "phase-shift'ng" device.

The rules are:

(1) $x_i \longrightarrow t_i t_i$, $0_i \longrightarrow 0_i'$, $1_i \longrightarrow 1_i' 1_i'$, $y_i \longrightarrow u_i u_i'$,

(2) $t_i \longrightarrow x_i' x_i''$,

(3) $0_i' \longrightarrow x_i' 0_i'' 0_i''$, $1_i' \longrightarrow 1_i'' 1_i''$, $u_i \longrightarrow y_i' y_i''$, $u_i' \longrightarrow y_i'' y_i'$,

(4) $x_i' \longrightarrow x_k x_k$, $x_i'' \longrightarrow x_{i+1} x_{i+1}$, $0_i'' \longrightarrow 0_{i+1} 0_{i+1}$, $1_i'' \longrightarrow 1_k 1_k$,

$y_i' \longrightarrow y_k y_k$, $y_i'' \longrightarrow y_{i+1} y_{i+1}$.

Deductions.

When $s = 0$, $(x_i x_i)^m 0_i 0_i (y_i y_i)^n$,

by (1), $\longrightarrow (t_i t_i)^m 0_i' (u_i u_i')^n$,

by (2), $\longrightarrow 0_i' u_i (u_i' u_i)^{n-1} u_i' x_i' (x_i'' x_i')^{m-1} x_i''$,

by (3), $\longrightarrow (x_i'' x_i')^m 0_i'' 0_i'' (y_i'' y_i')^n$,

by (4), $\longrightarrow (x_{i+1} x_{i+1})^m 0_{i+1} 0_{i+1} (y_{i+1} y_{i+1})^n$.

When $s = 1$,

ever $k > n$. Whenever we get to $n+1$ from q_n or by $C(n+1)$, the machine is to stop. We add simply a new symbol h and the rules:

$$x_{n+1} \longrightarrow h, \quad s_{n+1} \longrightarrow h, \quad y_{n+1} \longrightarrow h, \quad h \longrightarrow h.$$

This completes the proof of Theorem 5.

5.5 *Monogenic normal system.*

A normal system is any set of rules

$$N_i: \quad B_i \longrightarrow E_i$$

such that, for each given string. if it is B_iP, it becomes PE_i by the rule. It is monogenic if, for $i \neq j$, B_i is never B_j or B_j followed by some string. A different definition of monogenic system would be that any string can be broken up in at most one way into the B_i's. The two definitions are not quite the same, since the latter includes more. For example, if the alphabet is $\{0,1\}$, and B_1 is 00, B_2 is 001, B_3 is 11. Given any string, when we encounter 001, we have a question of taking it to be $B_1 1$ or B_2, but this can be settled by determining whether there are an even or an odd number of consecutive 1's immediately following 00. For our purpose, it is convenient to use the narrower definition.

It is very easy to use the SS machines to give a new proof of the known fact that there are monogenic normal systems with an unsolvable halting problem. Thus, as before, take a universal SS machine with n instructions. The rules for the corresponding monogenic normal system are simply:

(1) $\qquad\qquad\qquad 0 \longrightarrow 0.$

(2) $\qquad\qquad\qquad 1 \longrightarrow 1.$

For each q_i which is P_0:

(3) $\qquad\qquad\qquad b_i \longrightarrow b_{i+1}.$

(4) $\qquad\qquad\qquad e_i \longrightarrow 0e_{i+1}.$

For each q_i which is P_1:

(5) $\qquad\qquad\qquad b_i \longrightarrow b_{i+1}.$

(6) $\qquad\qquad\qquad e_i \longrightarrow 1e_{i+1}.$

For each q_i which is $SD(k)$:

(7) $b_i 0 \longrightarrow e_{i+1} b_{i+1}.$

(8) $b_i 1 \longrightarrow e_k b_k.$

(9) $e_i e_{i+1} \longrightarrow e_{i+1}.$

(10) $e_i e_k \longrightarrow e_k.$

Observe that if we reverse the arrows in the above rules, we do not get a monogenic system. We may modify (7)—(10) to read: $b_i 0 \longrightarrow e_i' b_{i+1}$, $b_i 1 \longrightarrow e_i'' b_k$, $e_i e_i' \longrightarrow e_{i+1}$, $e_i e_i'' \longrightarrow e_k$. Nevertheless, we would still have 0 at the right hand side of (1), $0 e_{i+1}$ at that of (4), 1 in (2), $1 e_{i+1}$ in (6); in fact e_{i+1} in (4) or (6) might be the same as e_k in (10).

INDEX

admissible ordinal   8, 117
—   set   9, 117
algorithm   57, 198, 201, 206, 207, 213
analytic set   117
analytical hierarchy   116, 163
arithmetical hierarchy   8, 115
—   partition   24
—   predicate   21
—   set   19, 114, 145, 149
Artin's conjecture   83
axiom of choice   158
— — determinacy   55, 156, 164
— — infinity   68
— — reducibility   125
— — replacement   164
— — separation   164
axiomatic method   2, 3
—   set theory   120
Boolean algebra   1
—   connective   13
—   expression   53
Boolean-valued models   138
— — set   138
Borel determinacy   164
—   game   164
—   set   138, 141
Burnside's problem   57
categorical   79, 80
cofinality   140
combinatorial system   257
combinatorics   137
comeager set   151, 152
compact cardinal   141
compactness theorem   15
completeness   14, 69, 71, 76, 88

complexity of compu-
tations and proofs   53
comprehension axiom   130
computability   206
consistency   21, 69, 71
constructibility   121, 134
constructible set   117, 124, 128, 130
constructivism   161
continuum hypothesis   5, 120, 127
—   problem   52, 135, 121, 127
countable anti-chain   154
—   model   5, 15, 128
—   ordinal   5, 129, 158
creative set   113
cut-elimination   160
decidability   68, 108, 109
decision problem   65, 186, 254
definability   124, 144, 207
degree of unsolvability   8, 113, 114
derivability problem   259, 262
Diophantine equation   9
domino problem   249, 254, 257
—   set   194, 195
—   type   187, 250, 251
dominoes   186, 187
Euler path   61
fan theorem   160
feasible   94, 95
Fermat's conjecture   51, 247
finitely axiomatizable   74, 78
finite-state machine   220, 222
first order logic   13, 63, 186, 254
— — Peano arith-
metic   18
— — theory   17
forcing   121, 141, 146
formal system   3, 11, 13, 15

formalization 11, 88

four-color theorem 42

general recursive func-
tion 6

generalized continuum
hypothesis 7

generic set 152

giant term 103, 104

Gödel's incompleteness
theorems 19

Goldbach's conjecture 113, 247

grid problem 103

Hamilton path 61

halting problem 247, 249, 250,
258, 265

hierarchy 112, 116

Hilbert's program 7, 156

— tenth problem 9

homogeneous sets 25, 26

impredicative definition 124, 158

inaccessible cardinal 137

independence 69, 135, 136, 153

indiscernibles 83

infinite state machine 220

infinity lemma 183, 186, 193

lag system 258

large cardinal axiom 163

Liar paradox 6

logical nets 223

Löwenheim-Skolem
theorem 15, 16, 77

meager set 151, 152

measurable cardinal 137, 164

measures of complexity 65

metamathematics 157

model 65

— theory 3, 73

monogenic normal
system 269

nim 176, 177, 198

nondeterministic 97, 98

nonstandard analysis 77

— model 2

occupancy problem 103

partial recursive func-
tion 113

Peano arithmetic 12, 17, 18, 74, 121

polynomial time 9, 96, 108

predicate calculus 2, 3, 13, 15

predicative 124

Presburger arithmetic 109

primitive recursive
function 94

principle of determi-
nacy 210

— — finiteness 210

printing problem 247

projective determinacy 163

— hierarchy 141, 163

— set 141, 163

proof theory 157

propositional calculus 1

provability 65, 207

quadratic time 96

quasilinear time 108

ramified hierarchy 7, 129, 137

— type theory 5

Ramsey's theorem 25, 27, 83, 84

rank model 126

recursion theory 4, 112

recursive function 112, 113

— set 112, 115, 117

recursively axiomati-
zable 74, 116

— enumerable
set 57, 242, 243

reducibility 112, 113

reflection principle 131

regularity property 163, 164

relative consistency 22, 130, 133

Richard's paradox 6

satisfiability 65, 71
second order theory 17
semantics 4
sequential machine 223
set theory 2, 3, 119
simple set 113
— theory of types 160
syntax 4
tag system 258
tautology problem 96, 97, 101
Thue sequence 56, 179, 180, 181, 186
tiling problem 110, 186, 249
torus 187, 188
Turing computability 213, 216
— machine 9, 57, 97, 186, 215, 220, 227
— reducible 113
ultrafilter 81
ultraproduct 80
undecidable 60
uniform halting problem 247
universal halting problem 245
unsolvability 110
validity 63, 65
vicious-circle principle 123, 125
word problem for group 57
Zermelo set theory 164